Ex-BR DIESELS IN INDUSTRY

Full details of those British Rail diesel locomotives
below 1000hp sold for industrial service and
preservation - past and present - at home and abroad

by
Adrian Booth

HANDBOOK 7BRD
INDUSTRIAL RAILWAY SOCIETY 2011

7BRD

ISBN: 978 1 901556 76 6

Seventh Edition
© Industrial Railway Society
26 Great Bank Road, Herringthorpe,
Rotherham, South Yorkshire, S65 3BT

www.irsociety.co.uk

Front cover photograph: D2953 is a historic locomotive as regards the BRD series for, when Thames Matex Ltd of West Thurrock acquired it in June 1966, it became the first ex-BR diesel shunter to enter industrial service. It was subsequently sold for preservation, and is seen here at Peak Rail, Rowsley, on 10th May 2008. (Brian Cuttell)

Produced for the IRS by Print Rite, Freeland, Witney, Oxon. 01993 881662

All rights reserved. No portion of this publication may be reproduced, stored in a retrieval system, or transmitted in any form or by any means without prior permission in writing from the Industrial Railway Society. Within the UK, exceptions are allowed in respect of any fair dealing for the purpose of private research or private study or criticism or review, as permitted under the Copyright, Designs and Patents Act 1988.

INTRODUCTION

The first edition of this book was published in 1972, being conceived and edited by the late Eric Tonks, then the President of the Industrial Railway Society. In the mid-1970s, at Eric's request, I took over his records and have maintained and updated them to this day. The second edition of BRD followed in 1981, the third in 1987, the fourth in 1991, the fifth in 2000, and the 6th in 2007.

The contents of the first four books were fairly straightforward, simply listing ex-British Railways diesel shunters working at industrial premises, together with those which had passed from industrial use to preservation. The handful of shunters that went directly to preservation were also listed, as they were akin to their industrial cousins. The information was compiled using notes and observations provided by members of the Society. Since publication of the fourth edition, however, there were massive changes in the British railway industry. This led to much thought as regards the contents of the fifth and sixth issues, in particular the effects of privatisation of main line railways, for it could be argued that ALL diesel shunters owned by privatised companies should then be included. Such a listing would have been impractical, however, and outside the scope of this series. A decision was taken (when preparing 5BRD) to include only shunters which worked at the privatised companies' workshops, and which were hired by them. These policies were continued in the sixth edition, and are applied now in this, the seventh edition.

The last two decades have seen many mergers, takeovers, and restructuring of companies, together with a move towards adopting 'modern' style trendy names. A decision has been made to show company names as used at the time locomotives were acquired, and not to record all the subsequent name changes. Many private individuals have purchased locomotives for preservation, and these are simply shown under the name of their home preservation site, as it would be difficult to ascertain, update, and show ownership of these locomotives. An exception to this rule is Harry Needle who owns a large number of industrial and ex-BR locomotives and who formed the 'Harry Needle Railroad Company'. His locomotives are often kept at preservation sites, and herein are indicated by (HNRC) immediately after the site name. It has been decided to include only locomotives rated at up to 999hp, and to exclude the much larger locomotives, which in no way can be considered as 'industrials'.

I would like to thank all IRS members who have sent me observations, information and photographs. I must, however, give special mention to Alex Betteney, Brian Cuttell and Bob Darvill who have been by far my most regular correspondents and have contributed much information. For this 7th edition, a special effort has been made to update the text as much as possible, and major thanks are due to Alex Betteney (who very carefully and thoroughly checked my manuscript and made very many useful suggestions). Thanks also to Andrew Smith (who trawled through files of archived members letters sorting out many additional 'exact dates' for locomotive movements) and to Brian Cuttell (who located numerous locomotive withdrawal dates/depot codes). My thanks also to all the individuals, NCB officials, and private companies who responded to my requests for information during the thirty-six years that I have maintained these records. The locomotive listings herein incorporate all information received up to 18th August 2011.

Adrian Booth
Rotherham, 18th August 2011

EXPLANATORY NOTES

Layout

The book is divided into 28 sections, each devoted to one class, and brief details appropriate to each are given. The engines, etc, apply to locomotives as built, and not necessarily as when sold out of service. All are standard gauge unless otherwise stated. Sections 1 to 28 are followed by Appendices A, B and C. Each locomotive has eight columns of information, thus:

| BR number | Builder | Works number | Year built | Last depot | Date wdn | P/F | Title |

These will be simple to understand, but the following notes are pertinent:-

BR number

The 1957 numbering scheme (with the 'D' prefix) has been used to dictate the order of the classes in the book, although the later TOPS computer numbers (where allocated) are also given.

Builder

Many locomotives were built at British Railways' own workshops at Ashford, Crewe, Darlington, Derby, Doncaster, Horwich and Swindon, and these are shown where appropriate. Locomotives which were constructed at private companies' workshops are indicated by standard IRS abbreviations, as under. In the case of Drewry locomotives two numbers are given: this company was basically a sales organisation, which did not erect locomotives but rather sub-contracted the construction work, in practice to RSH or VF. Both organisations' works numbers are given in these cases.

AB	Andrew Barclay, Sons & Co Ltd, Kilmarnock
CE	Clayton Equipment Co Ltd, Hatton, Derby
DC	Drewry Car Co Ltd, London
EE	English Electric Co Ltd, London
HC	Hudswell, Clarke & Co Ltd, Leeds
HE	Hunslet Engine Co Ltd, Leeds
NB	North British Locomotive Co Ltd, Glasgow
RH	Ruston & Hornsby Ltd, Lincoln
RSHD	*Robert Stephenson & Hawthorns Ltd, Darlington
RSHN	*Robert Stephenson & Hawthorns Ltd, Newcastle upon Tyne
VF	*Vulcan Foundry Ltd, Newton-le-Willows, Lancashire
YE	Yorkshire Engine Co Ltd, Sheffield

*(latterly owned by English Electric)

Works number

This only applies to locomotives built by private companies. Those locomotives constructed at British Railways workshops were not allocated individual works numbers.

Year built

The year quoted is that in which the locomotive was officially added to British Railways stock, or the worksplate date if this is known to be different.

Last depot

During their 'main line' life, all the locomotives listed in this book were allocated to various British Railways motive power depots. Up to 6th May 1973 these depots were identified by an official code that comprised a number and letter; after this date a new system of two-letter codes was adopted. The list below shows all the depots and workshops from which locomotives listed in this book were withdrawn. The original and new codes for all of these depots are shown below although, in most cases, the older style code is used in the listings.

1A	WN	Willesden (London)	32A	NR	Norwich Thorpe
1E	BY	Bletchley		NC	Norwich Crown Point
2E	SY	Saltley (Birmingham)	34E	PB	New England (Peterborough)
2F	BS	Bescot (Walsall)	36A	DR	Doncaster
5A	CD	Crewe Diesel	36C	FH	Frodingham (Scunthorpe)
6A	CH	Chester	40A	LN	Lincoln
6G	LJ	Llandudno Junction	40B	IM	Immingham
8C	—	Speke Junction (Liverpool)	41A	TI	Tinsley (Sheffield)
8F	SP	Springs Branch (Wigan)	41J	SB	Shirebrook West
8H	BC	Mollington Street (Birkenhead)	50B	—	Dairycoates (Hull)
			50C	BG	Botanic Gardens (Hull)
	BD	Birkenhead North	50D	GO	Goole
8J	AN	Allerton (Liverpool)	51A	DN	Darlington
9A	LO	Longsight (Manchester)	51L	TE	Thornaby (Middlesbrough)
9D	NH	Newton Heath (Manchester)	52A	GD	Gateshead
10D	LH	Lostock Hall (Preston)	52B	HT	Heaton (Newcastle)
12A	KD	Kingmoor Diesel (Carlisle)	—	TY	Tyne Yard (Gateshead)
12B	CL	Upperby (Carlisle)	55A	HO	Holbeck (Leeds)
12C	BW	Barrow	55B	YK	York
15A	LR	Leicester Midland	—	YC	York Clifton (from Oct. 1983)
16A	TO	Toton (Nottingham)	55C	HM	Healey Mills (Wakefield)
16B	—	Colwick (Nottingham)	55F	HS	Hammerton Street (Bradford)
16C	DY	Derby	55G	KY	Knottingley
16F	BU	Burton-on-Trent	55H	NL	Neville Hill (Leeds)
30A	SF	Stratford (London)	60A	IS	Inverness
30E	CR	Colchester	62A	TJ	Thornton Junction (Kirkcaldy)
31A	CA	Cambridge	62C	DT	Townhill (Dunfermline)
31B	MR	March	64B	HA	Haymarket (Edinburgh)

64H	—	Leith Central (Edinburgh)	82C	SW	Swindon	
—	EC	Craigentinny (Edinburgh)	83B	—	Taunton	
65A	ED	Eastfield (Glasgow)	84A	LA	Laira (Plymouth)	
66A	PO	Polmadie (Glasgow)	85A	WS	Worcester	
66B	ML	Motherwell	85B	GL	Gloucester	
67C	AY	Ayr	86A	CF	Canton (Cardiff)	
70D	EH	Eastleigh	86E	ST	Severn Tunnel Junction	
70F	BM	Bournemouth West	87B	MG	Margam	
70H	RY	Ryde (Isle of Wight)	87E	LE	Landore (Swansea)	
73C	HG	Hither Green (London)		ZC	Crewe Works	
73F	AF	Chart Leacon (Ashford)		ZF	Doncaster Works	
75A	BI	Brighton		ZG	Eastleigh Works	
75C	SU	Selhurst (London)		ZH	St Rollox (Glasgow) Works	
81A	OC	Old Oak Common (London)		ZJ	Horwich Works	
81D	RG	Reading		ZI	Ilford Works	
81F	OX	Oxford		ZL	Swindon Works	
82A	BR	Bath Road (Bristol)		ZN	Wolverton Works	
82B	PM	St Philips Marsh (Bristol)				

The following are codes that have been adopted for this book only:-
BSD   Beeston Sleeper Depot, near Nottingham
CJ   Chesterton Junction Permanent Way Materials Depot, Cambridge
HHC   Hall Hills Creosoting Depot, Boston
LSD   Lowestoft Sleeper Depot, Suffolk
MQ   Meldon Quarry, near Okehampton, Devon
NTW   Not Technically Withdrawn
RSD   Reading Signal Depot

Date withdrawn

This is the official date of withdrawal from British Railways stock. Official BR information often quoted a locomotive being withdrawn during a period, as for example "7-12-86 to 3-1-87". In such cases the last date has been taken, and the example quoted would be shown as 1/87 in the following tables. As regards post-BR days, it has often proved difficult to find information as regards withdrawal dates, and there are several locomotives for which this is not shown.

P/F and Title

These columns indicate whether the stated number and/or name is the Present (P) or Former (F) one carried by the locomotive. NPT indicates 'No Present Title', meaning the locomotive presently carries neither name nor number. Former titles given for scrapped locomotives are those carried by the locomotive at the time it was cut up. NFT indicates 'No former title', meaning the locomotive carried neither name nor number at the time it was scrapped.

Notes

These follow on from the basic locomotive data. The first industrial user (or preservation society) is stated, together with the date the locomotive arrived on site. In certain cases the move was protracted, for example due to a 'hot box', and these are indicated by a period, such as 'June to September 1969'. This means the locomotive left its main line depot in June 1969, but did not arrive at its destination until September 1969. Locomotives were moved from their last main line depot unless otherwise stated. All known subsequent movements and disposals are given, with the date when known. Certain abbreviations are used:

APCM	Associated Portland Cement Manufacturers Ltd
BHESS	Barrow Hill Engine Shed Society
BR	British Railways
BSC	British Steel Corporation
CEGB	Central Electricity Generating Board
DBS	Deutsch Bahn Schenker
EWS	English Welsh & Scottish Railway Ltd
GBRf	GB Railfreight Ltd
GNER	Great North Eastern Railway Ltd
HNRC	Harry Needle Railroad Company Ltd
NCB	National Coal Board
NCBOE	National Coal Board Opencast Executive
NSF	National Smokeless Fuels Ltd

LOCOMOTIVE LISTINGS

SECTION 1

British Railways built 0-6-0 diesel mechanical locomotives, numbered D2000-D2199, and D2370-D2399, introduced 1957. Fitted with a Gardner 8L3 engine developing 204bhp at 1200rpm, a five speed gearbox, and driving wheels of 3ft 7in diameter. Later classified TOPS Class 03.

D2012   Swindon   1958   31B   12/75   F   03012/ F135L
03012   to A. King & Sons Ltd, Snailwell, Cambridgeshire, 28th July 1976; scrapped 28/29th January 1991.

D2018   Swindon   1958   32A   11/75   P   No.2/ 600
03018   to George Cohen, Sons & Co Ltd, Cransley, Northamptonshire, 29th April 1976; to 600 Fragmentisers Ltd, Willesden, London, October 1980; to Mayer Parry Ltd, Snailwell, 17th July 1995; to South Yorkshire Railway Preservation Society (HNRC), Meadowhall, Sheffield, 11th November 1998; to Lavender Line, Isfield, about June 2001; to Mangapps Farm Railway Museum, Burnham-on-Crouch, Essex, 1st March 2004.

D2020   Swindon   1958   32A   12/75   P   03020/ F134L
03020   to A. King & Sons Ltd, Snailwell, Cambridgeshire, July 1976; to South Yorkshire Railway Preservation Society (HNRC), Meadowhall, Sheffield, 9th November 1995; to Lavender Line, Isfield, for storage, 27th July 2001; to Sonic Rail Ltd, Burnham-on-Crouch, Essex, 13th December 2010.

D2022   Swindon   1958   52A   11/82   P   2022
03022   to Swindon & Cricklade Railway (despatched from BR Swindon Works), 18th November 1983; to Coopers (Metals) Ltd, Swindon, on hire, early October 1989; returned to Swindon & Cricklade Railway, summer 1995.

D2023   Swindon   1958   40A   7/71   P   D2023
to Tees & Hartlepool Port Authority, Middlesbrough Docks, July 1972; to T&HPA, Grangetown Docks, 15th September 1980; to Kent & East Sussex Railway, Tenterden, 14th August 1983.

D2024   Swindon   1958   40A   7/71   P   4
to Tees & Hartlepool Port Authority, Middlesbrough Docks, July 1972; to T&HPA, Grangetown Docks, 15th September 1980; to Kent & East Sussex Railway, Tenterden, 4th September 1983.

D2027   Swindon   1958   30E   1/76   P   NPT
03027   to Shipbreakers (Queenborough) Ltd, Kent, July 1976; to South Yorkshire Railway Preservation Society, Meadowhall, Sheffield, 9th February 1991; to Knights of Old Ltd, Old, Northamptonshire, 28th September 1993; to Peak Rail, Darley Dale, 2nd January 1997.

D2037   Swindon   1959   32A   9/76   P   NPT
03037   to Hargreaves Industrial Services Ltd, NCBOE British Oak Disposal Point, Crigglestone, October 1977; to NCBOE West Hallam Disposal Point, Mapperley, November 1983; to NCBOE British Oak Disposal Point, Crigglestone, July 1984; to NCBOE Oxcroft Disposal Point, Clowne, November 1988; to South Yorkshire Railway Preservation Society (HNRC), Meadowhall, Sheffield, 18th July 1995; to Lavender Line, Isfield, about June 2001; to Peak Rail, Rowsley, 18th May 2004.

D2041   Swindon   1959   75C   2/70   P   D2041
to CEGB Richborough, February 1970: to CEGB Rye House, Hoddesdon, Hertfordshire, March 1971; to CEGB Barking, about May 1971; to CEGB Rye House, August 1974; to Colne Valley Railway, Essex, 15th January 1981.

D2046   Doncaster   1958   51L   10/71   P   NPT
Rebuilt by Hunslet (6644 of 1967); to Gulf Oil Co Ltd, Waterston, Milford Haven, Pembrokeshire, May 1972; to BR Canton Depot, Cardiff, for tyre turning, 16th May 1975; returned to Gulf Oil; to BR Canton Depot, Cardiff, for tyre turning, 5th November 1980; returned to Gulf Oil; to Petro-Plus, Waterston, Milford Haven, Pembrokeshire, with site, 1998; to Beavor Locomotive Company, Dowlais, Merthyr Tydfil, January 2005; to Moveright International, Wishaw, Warwickshire, about March 2006; to Plym Valley Railway, Marsh Mills, Plymouth, 28th May 2008.

D2049   Doncaster   1958   50D   8/71   F   D2049
to Hargreaves Industrial Services Ltd, NCBOE Bowers Row Disposal Point, Astley, West Yorkshire, by 12th May 1974; to NCBOE British Oak Disposal Point, Crigglestone, by 6th January 1975; to NCBOE West Hallam Disposal Point, Mapperley, Derbyshire, March 1978; to NCBOE British Oak Disposal Point, Crigglestone, 28th January 1985; scrapped on site by Wath Skip Hire Ltd of Rotherham, November 1985.

D2051   Doncaster   1959   30E   12/72   P   D2051
to Ford Motor Co Ltd, Dagenham, London, about September 1973; to BR Swindon Works, for rebuild, 3rd October 1977; returned to Dagenham, 23rd February 1978; to Rother Valley Railway, Robertsbridge, East Sussex, 6th November 1997; to North Norfolk Railway, Sheringham, by 28th September 2000.

D2054   Doncaster   1959   55B   11/72   F   CENTA
to Chair Centre Ltd, Derby, October 1973; to British Industrial Sand Ltd, Middleton Towers, Norfolk, 13th July 1979; to C.F. Booth Ltd, Rotherham, May 1982; scrapped, September 1982.

D2057   Doncaster   1959   51L   10/71   F   No.1
Rebuilt by Hunslet (6645 of 1967); to NCB Grimethorpe Colliery, Barnsley, 19th September 1972; to C.F. Booth Ltd, Rotherham, 25th March 1986; scrapped, April 1986.

D2059   Doncaster   1959   NC   7/87   P   D2059/ EDWARD
03059   to Isle of Wight Steam Railway, Havenstreet (despatched from BR Colchester), 4th November 1988; to Island Line, on hire, 16th March 2002; returned to IoWSR, 19th March 2002.

D2062   Doncaster   1959   32A   12/80   P   D2062
03062   to Dean Forest Railway Society Ltd (despatched from BR Swindon Works), 30th September 1982; to Yeovil Country Railway, Yeovil Junction, about September 1995; to East Lancashire Railway, Bury, 13th October 1997; to Metrolink, Manchester, on hire, (upgrade contract), 26th May 2007; returned to East Lancashire Railway, about 12th September 2007.

D2063   Doncaster   1959   52A   12/87   P   D2063
03063   to Colne Valley Railway, Essex, 11th November 1988; to East Anglian Railway Museum, Essex, March 1995; to North Norfolk Railway, Sheringham, 23rd February 2000.

D2066   Doncaster   1959   52A   1/88   P   03066
03066   despatched from BR Gateshead, 28th June 1988; to Horwich Foundry Ltd, Horwich, 30th June 1988; to South Yorkshire Railway Preservation Society, Meadowhall, Sheffield, 5th April 1991; to Barrow Hill Engine Shed Society, Staveley, 23rd March 1999.

D2069   Doncaster   1959   52A   12/83   P   2069
03069   to Vic Berry Company, Leicester, 4th January 1984; auctioned by Walker Walter Hanson of Nottingham, 1st August 1991, and sold for £9,500; to the Gloucestershire Warwickshire Railway Society, Toddington, 2nd August 1991; to Wabtec, Doncaster (for open day), July 2003; returned to Toddington, May 2004; to Crewe Works, for open day, September 2005; returned to Toddington, September 2005; on low-loader at Strensham North, M6 motorway (destination not known), 6th January 2006; returned to Toddington by May 2006.

D2070   Doncaster   1959   51L   11/71   F   D2070
to Shipbreakers (Queenborough) Ltd, Kent, June 1972; to South Yorkshire Railway Preservation Society, Meadowhall, Sheffield, 5th October 1990; to Churnet Valley Railway, Cheddleton, 11th September 1993; to Cotswold Rail, Moreton in Marsh, 2001; scrapped, July 2001.

D2072   Doncaster   1959   51A   3/81   P   D2072
03072   to Lakeside & Haverthwaite Railway Co Ltd, Cumbria, August 1981.

D2073   Doncaster   1959   BD   5/89   P   03073
03073   to BR Chester for storage, 30th March 1989; despatched from BR Chester, 17th January 1991; arrived at Railway Age, Crewe, 18th February 1991; displayed at Crewe Depot open day, 12th October 1991; to East Lancashire Railway, Bury, 1st October 1992; returned to Crewe, 8th October 1992; displayed at Crewe Railfare, 21st August 1994; to Greater Manchester Metro Ltd, on hire, 19th August 1995; returned to Crewe, 16th September 1995; to Greater Manchester Metro Ltd, on hire, 26th September 1996; returned to Crewe, 8th October 1996; to Greater Manchester Metro Ltd, on hire, 5th September 1997; returned to Crewe, 22nd September 1997; to Greater Manchester Metro Ltd, on hire, 12th March 1999; returned to Crewe, 30th March 1999; to Greater Manchester Metro Ltd, on hire, 9th January 2001; returned to Crewe, by 3rd March 2001; to Crewe Works, for open day, 3rd September 2005; returned to Railway Age, early 2006.

D2078   Doncaster   1959   52A   1/88   P   D2078
03078   to Stephenson Railway Museum, Chirton, Tyne & Wear, 11th May 1988.

D2079   Doncaster   1960   70H   6/98   P   03079
03079   sold by Island Line, Isle of Wight (BR's successor), and moved to Great Yorkshire Railway Preservation Society, Murton, York, 6th June 1998.

D2081   Doncaster   1960   31B   12/80   P   03081/LUCIE
03081   ex Appendix C; to Mangapps Farm Railway Museum, Burnham-on-Crouch, Essex, 8th March 2004.

**D2084   Doncaster   1959   NC   6/87   P   03084/HELEN-
                                                                          LOUISE**
03084   to Knights of Old Ltd, Old, Northamptonshire (despatched from 31B March Depot), 26th January 1992; to Peak Rail, Darley Dale, Matlock, 4th January 1997; to

Ecclesbourne Valley Railway, Wirksworth, 16th March 2005; to LH Group, Barton under Needwood, for repairs, 7th July 2006; returned to EVR, Wirksworth, 21st March 2007; to Lincolnshire Wolds Railway, Ludborough, 15th June 2009; to West Coast Railway, Carnforth, 11th January 2011.

D2089   Doncaster   1960   NC   12/87   P   03089
03089   to British Sugar Corporation Ltd, Woodston Factory, Peterborough, for storage (despatched from 31B March Depot), 2nd September 1988; to Nene Valley Railway, about September 1990; to Mangapps Farm Railway Museum, Burnham-on-Crouch, Essex, 3rd October 1991.

D2090   Doncaster   1960   55B   7/76   P   03090
03090   to National Railway Museum, York, August 1976; to National Railway Museum, Shildon, 10th June 2004.

D2093   Doncaster   1960   51L   10/71   F   No.2
Rebuilt by Hunslet (6643 of 1967); to NCB Grimethorpe Colliery, Barnsley, 19th September 1972; to C.F. Booth Ltd, Rotherham, 26th March 1986; scrapped, April 1986.

D2094   Doncaster   1960   52A   1/88   P   D2094
03094   despatched from BR Gateshead, 28th June 1988; to Horwich Foundry Ltd, Horwich, 30th June 1988; to South Yorkshire Railway Preservation Society, Meadowhall, Sheffield, 6th April 1991; to Barrow Hill Engine Shed Society, Staveley, 23rd March 1999; to Cambrian Railway Trust, Llynclys, Oswestry, 2nd June 2005; to Royal Deeside Railway, 13th May 2010.

D2099   Doncaster   1960   51L   2/76   P   03099
03099   to National Smokeless Fuels Ltd, Fishburn Coking Plant, County Durham, July 1976; to Monkton Coking Plant, Hebburn, 26th February 1981; to South Yorkshire Railway Preservation Society, Meadowhall, Sheffield, about September 1992; to Exhibition of Steam and Speed, Doncaster, for display, July 2000; returned to SYRPS, 17th July 2000; to Peak Rail, Rowsley, March 2002.

D2112   Doncaster   1960   NC   7/87   P   D2112
03112   to British Sugar Corporation Ltd, Woodston Factory, Peterborough, for storage (despatched from 31B March Depot), 2nd September 1988; to Nene Valley Railway, about September 1990; to Port of Boston, Boston, on hire, 20th January 1992; returned to NVR, 29th January 1992; to Port of Boston, 11th September 1998; to Nene Valley Railway, for display, 5th October 2006, returned to Port of Boston, 8th October 2006.

D2113   Doncaster   1960   55B   8/75   P   NPT
03113   to Gulf Oil Co Ltd, Waterston, Milford Haven, 17th May 1976; to Maritime & Heritage Museum, Milford Haven, 17th October 1991; to Peak Rail, Rowsley, 6th to 16th November 2002.

D2114   Swindon   1959   82C   5/68   F   D2114
to Bird's Commercial Motors Ltd, Long Marston, Worcestershire (despatched from 85A Worcester Depot), August 1968; scrapped, January 1975.

D2117   Swindon   1959   8F   10/71   P   D2117
to Lakeside & Haverthwaite Railway, Cumbria, 24th April 1972.

D2118   Swindon   1959   12C   6/72   P   D2118
to Anglian Building Products Ltd, Lenwade, Norfolk, about August 1973; to Costain Dow-Mac Ltd, Tallington, Stamford, 7th December 1993; to South Yorkshire Railway Preservation Society (HNRC), Meadowhall, Sheffield, 25th March 1996; to Rutland Railway, on loan, 20th March 2001; to Peak Rail, Rowsley, 31st March 2005; to Great Central Railway, Ruddington, 28th January 2011.

D2119   Swindon   1959   87E   2/86   P   03119
03119   to A.E. Knill Ltd, Barry, by 18th November 1986; to Dean Forest Railway, 19th December 1986; to Rail & Maritime Engineering, Thingley Junction, Chippenham, for storage, about March 1995; to West Somerset Railway, Minehead, 28th March 1996.

D2120   Swindon   1959   87E   2/86   P   D2120
03120   to Sir W.H. McAlpine, The Fawley Hill Railway, Buckinghamshire, 18th December 1986.

D2122   Swindon   1959   82A   11/72   F   D2122
to John Cashmore Ltd, Newport, 1974; to Briton Ferry Steel Co Ltd, Glamorgan, in a dismantled state, 13th August 1974; used for spares; remains scrapped by August 1976.

D2123   Swindon   1959   82C   12/68   F   NFT
to Bird's Commercial Motors Ltd, Long Marston, Worcestershire, June 1969; to Bird's, Cardiff, April 1970, where converted to a stationary generator; to Derby Power Station, Full Street, Derby, by 8th November 1970, where used as a generator on demolition contract; to Bird's, Stapleton Road, Bristol, for storage, 1971; later used as a stationary generator; still there in November 1978, in a dismantled state, but scrapped shortly afterwards.

D2125   Swindon   1959   82C   12/68   F   NFT
to Bird's (Swansea) Ltd, 40 Acre Site, Cardiff, May 1969; scrapped, June 1976.

D2128   Swindon   1960   82A   7/76   P   D2128/HEIDI
03128   to Bird's Commercial Motors Ltd, Long Marston, October 1976; resold for export, minus its engine; shipped from Harwich Docks, December 1976; to Zeebouw-Zeezand (numbered 6G2), for Zeebrugge Port Authority harbour extension contract; at Gent shed for wheel turning and fitting with Deutz V12 engine, March 1989; to Stoomcentrum, Maldegem, May 1989; to Peak Rail, Darley Dale, Matlock, 29th June 1993; to Nottingham Sleeper Co Ltd, Elkesley, Retford, 24th April 1996; to Cotswold Rail, Moreton in Marsh, about September 2000; to Dean Forest Railway, about July 2002; to Blast Furnace Sidings, Corus, Scunthorpe, for storage, 8th July 2008; to Appleby Frodingham Railway Society, Scunthorpe, August 2008.

D2132   Swindon   1960   82C   5/69   F   D2132/ LESLEY
to NCB Bestwood Colliery, Nottinghamshire, October 1970; to New Hucknall Colliery, July 1971; to Pye Hill Colliery, about 1981; to C.F. Booth Ltd, Rotherham, for scrap, 27th November 1984; scrapped, 28th/29th November 1984.

D2133   Swindon   1960   82A   7/69   P   D2133
to British Cellophane Ltd, Bridgwater, Somerset, July 1969; to West Somerset Railway, Minehead, 10th July 1996.

D2134   Swindon   1960   82A   7/76   P   03134
03134   to Bird's Commercial Motors Ltd, Long Marston, October 1976; resold for export, about January 1977; to Zeebouw-Zeezand (numbered 6G1), for Zeebrugge Port Authority harbour extension contract; at Gent shed for wheel turning and fitting with Deutz V12 engine, March 1989; to Stoomcentrum, Maldegem, May 1989; to South Yorkshire Railway Preservation Society, Meadowhall, Sheffield, 27th April 1995; to Royal Deeside Railway, Banchory, near Aberdeen, 12th December 2000.

D2138   Swindon   1960   82C   5/69   P   D2138
to NCB Bestwood Colliery, Nottinghamshire, 23rd October 1970; to Pye Hill Colliery, April 1971; to BR Toton for wheel turning, June 1978; returned to Pye Hill Colliery, 1978; to Midland Railway, Butterley, Derbyshire, 20th August 1985.

D2139   Swindon   1960   85A   5/68   P   NPT
to A.R. Adams & Son, Newport, about October 1968; used as hire loco (see Appendix A); sold to NSF Coed Ely Coking Plant, Tonyrefail, Mid Glamorgan by March 1971; to BR Canton Depot, Cardiff, for repairs, 29th September 1977; returned to Coed Ely, 1978; to BR Swindon Works, for repairs, 31st March 1981; returned to Coed Ely, 2nd March 1982; to Monkton Coking Plant, Hebburn, Tyne & Wear, December 1983; to South Yorkshire Railway Preservation Society (HNRC), Meadowhall, Sheffield, about September 1992; to Peak Rail, Rowsley, March 2002.

D2141   Swindon   1960   87E   7/85   P   03141
03141   to White Wagtail Ltd, Gun Range Farm, Shilton, near Coventry (despatched from 86E Severn Tunnel Junction Depot), 30th June 1986; to Cotswold Rail, Moreton in Marsh, about December 2000; to Dean Forest Railway, on hire, spring 2002; to Swansea Vale Railway, 19th June 2005; to Pontypool & Blaenavon Railway, 29th April 2008.

D2144   Swindon   1960   87E   2/86   P   2144/WESTERN
03144   WAGGONER
to MoD Long Marston, Worcestershire, 24th March 1987; to Grantham Barracks for display, June 1992; to MoD Bicester, 26th June 1992; to Yorkshire Engine Company, Long Marston, 11th December 1995; to 275 Squadron, MoD Bicester, 5th September 1996; to Wensleydale Railway, by 25th January 2004.

D2145   Swindon   1960   87E   7/85   P   03145
03145   to White Wagtail Ltd, Gun Range Farm, Shilton, near Coventry (despatched from BR Gloucester), 1986; to Cotswold Rail, Moreton in Marsh, about December 2000; to D2578 Locomotive Group, Moreton on Lugg, 6th August 2001.

D2146   Swindon   1961   87E   9/68   F   D2146
to Bird's Commercial Motors Ltd, Long Marston, Worcestershire (despatched from 82C Swindon Depot), June 1969; used in Army training exercise, MoD Long Marston, 16th October 1971; returned to Bird's, Long Marston; scrapped, July 1978.

D2148   Swindon   1960   55C   11/72 P   D2148
to Hargreaves Industrial Services Ltd, NCBOE Bowers Row Disposal Point, Astley, Yorkshire, August 1973; to Lindley Plant Ltd, NCBOE Gatewen Disposal Point, Denbighshire, September 1973; to Bowers Row, December 1973; following collision damage to its original cab, 03149's cab was purchased from BR Doncaster Works and fitted on site, January 1984; to Steamport Transport Museum, Southport, Merseyside, 14th March 1987; to Ribble Steam Railway, Preston, 17th April 1999; to EWS Crewe

Electric Depot, for tyre turning, 23rd January 2009; returned to Ribble, 15th February 2009.

D2150   Swindon   1960   55B   11/72   F   NFT
to British Salt Ltd, Middlewich, Cheshire, May 1973; to Telford Steam Railway, Shropshire, 13th April 2000; to Cotswold Rail, Moreton in Marsh, September 2000; used for spares; remains scrapped, July 2001.

D2152   Swindon   1960   87E   10/83   P   D2152
03152   to Swindon & Cricklade Railway (despatched from BR Swindon Works), 6th March 1986; to Swindon Heritage Centre, April 1988; displayed at Membury Services West (on the M4) from about April 1990; returned to Swindon Heritage Centre, November 1990; to Swindon & Cricklade Railway, about March 1993.

D2158   Swindon   1960   NC   6/87   P   D2158/MARGARET
03158   **ANN**
to Knights of Old Ltd, Old, Northamptonshire (despatched from 31B March Depot), 25th January 1992; to Peak Rail, Darley Dale, Matlock, 3rd January 1997; to Ecclesbourne Valley Railway, Wirksworth, 7th September 2004; to Lincolnshire Wolds Railway, Ludborough, 16th June 2009; to Great Central Railway, Ruddington, 12th August 2009.

D2162   Swindon   1960   BD   5/89   P   D2162
03162   purchased by Wirral Borough Council and moved for storage to BR Chester, 30th March 1989; to Llangollen Railway (despatched from BR Chester Depot), October 1989.

D2170   Swindon   1960   BD   5/89   P   03170
03170   to BR Chester for storage, 30th March 1989; to Otis Euro Trans Rail Ltd, Manchester (despatched from BR Chester), 24th July 1989; to Harry Needle Railroad Company, December 1999; to Fragonset, Derby, for certification, about April 2000; returned to Otis; to Barrow Hill Engine Shed Society, Staveley, 28th September 2000; to Battlefield Line, Shackerstone, 13th August 2001; to Epping & Ongar Railway, 17th September 2010.

D2176   Swindon   1961   ZC   5/68   F   D2176
to George Cohen, Sons & Co Ltd, Cransley, Northamptonshire, October 1968; scrapped, November 1971.

D2178   Swindon   1962   81F   9/69   P   D2178
to A.R. Adams & Son, Newport, Monmouthshire, February 1970; used as a hire loco (see Appendix A); sold to National Smokeless Fuels Ltd, Coed Ely Coking Plant, Tonyrefail, Mid Glamorgan, May 1974; to BR Swindon Works, for repairs, by July 1979; at BR Severn Tunnel Junction Depot on 1st November 1979; to Coed Ely, about February 1980; to Caerphilly Railway Preservation Society, Caerphilly, 12th November 1985; to Gwili Railway Company, Bronwydd Arms, 21st October 1996.

D2179   Swindon   1962   NTW   —-   P   03179/ CLIVE
03179   to West Anglia & Great Northern Company, Electric Maintenance Depot, Hornsey, 8th June 1998; to Nene Valley Railway, for gala, 29th February 2008; returned to First Capital Connect, Hornsey, about March 2008.

D2180   Swindon   1962   NC   3/84   P   03180
03180   to Mayer Newman Ltd, Snailwell, Newmarket, Cambridgeshire, 26th July 1984; to South Yorkshire Railway Preservation Society (HNRC) Meadowhall, Sheffield, South Yorkshire, 21st December 1991; to Battlefield Line, Shackerstone, 2nd August 2001.

D2181   Swindon   1962   87E   5/68   F   PRIDE OF GWENT
to A.R. Adams & Son, Newport, Monmouthshire (despatched from BR Worcester), 13th December 1968; used as a hire loco (see Appendix A); sold to Gwent Coal Distribution Centre, Newport, Monmouthshire, by August 1971; to Marple & Gillott Ltd, Sheffield, for scrap, December 1986; scrapped, January 1987.

D2182   Swindon   1962   87E   5/68   P   D2182
to A.R. Adams & Son, Newport, Monmouthshire (despatched from 85A Worcester Depot), 29th November 1968; used as a hire loco (see Appendix A); sold to Lindley Plant Ltd, Gatewen Disposal Point, Denbighshire, September 1973; to NCBOE Bennerley Disposal Point, 1981; to NCBOE Wentworth Stores, Rotherham, 18th March 1982; to Bennerley Disposal Point, 6th May 1983; to Coalfield Farm Disposal Point, Hugglescote, Leicestershire, July 1983; to Warwick District Council, Victoria Park, Leamington, 20th April 1986; to Gloucestershire Warwickshire Railway, Toddington, 11th January 1993.

D2184   Swindon   1962   87E   12/68   P   D2184
to Co-operative Wholesale Society Ltd, Coal Concentration Depot, Southend (despatched from 85A Worcester Depot), 20th August 1969; to Colne Valley Railway, Essex, 17th October 1986.

D2185   Swindon   1962   85A   12/68   F   D2185
to Bird's Commercial Motors Ltd, Long Marston, Worcestershire, 21st May 1969; to Abercarn Tinplate Works, on hire, April 1970; to Bird's, 40 Acre Site, Cardiff, early 1972; to Bird's, Long Marston, January 1978; scrapped, June 1978.

D2186   Swindon   1962   81F   9/69   F   D2186
to A.R. Adams & Son, Newport, Monmouthshire, 8th February 1970; used as a hire loco (see Appendix A); scrapped, January 1981.

D2187   Swindon   1961   82C   5/68   F   NFT
to Bird's Commercial Motors Ltd, Long Marston, Worcestershire (despatched from 85A Worcester Depot), September 1968; scrapped, June 1978.

D2188   Swindon   1961   83B   5/68   F   D2188
to Bird's Commercial Motors Ltd, Long Marston, Worcestershire (despatched from BR Swindon), September 1968; scrapped, February 1978.

D2189   Swindon   1961   6A   3/86   P   03189
03189   to Steamport, Southport (despatched from 31B March Depot), January 1992; to Ribble Steam Railway, Preston, 17th April 1999.

D2192   Swindon   1961   82C   1/69   P   D2192/ TRITON
to Dart Valley Railway, Devon, 25th August 1970; to Torbay & Dartmouth Railway, Paignton, 24th July 1977; to South Devon Railway Trust, Buckfastleigh, summer 1991; to Paignton and Dartmouth Steam Railway, by 21st June 1992.

D2193   Swindon   1961   82C   1/69   F   2
to A.R. Adams & Son, Newport, Monmouthshire (despatched from 85A Worcester Depot), October 1969; used as a hire loco (see Appendix A); scrapped, January 1981.

D2194   Swindon   1961   85A   9/68   F   D2194
to Bird's Commercial Motors Ltd, Long Marston, Worcestershire (despatched from 85A Worcester Depot), about March 1969; scrapped, July 1978.

D2195   Swindon   1961   82A   9/68   F   D10
to Llanelly Steel Co Ltd, Carmarthenshire, about May 1969 (sold per R.E. Trem Ltd, Finningley, Doncaster and despatched from 85A Worcester Depot); scrapped, September 1981.

D2196   Swindon   1961   8H   6/83   P   03196/ JOYCE
03196   to R.O. Hodgson Ltd, Carnforth, Lancashire, 15th June 1983; to Steamtown, Carnforth, by July 1992.

D2197   Swindon   1961   NC   6/87   P   03197
03197   to South Yorkshire Railway Preservation Society (HNRC), Meadowhall, Sheffield (despatched from 15A Leicester Depot), 25th October 1991; to Lavender Line, Isfield, for storage, 31st July 2001; to Sonic Rail Ltd, Burnham-on-Crouch, Essex, 13th December 2010.

D2199   Swindon   1961   12C   6/72   P   D2199
to BR Doncaster Works for overhaul, summer 1973; to NCB Rockingham Colliery, Birdwell, Barnsley, February 1974; to Barrow Colliery, Worsborough, Barnsley, about January 1979; to Houghton Main Colliery, Barnsley, about June 1979; to Royston Drift Mine, Barnsley, 14th August 1980; to Barrow Colliery, 8th July 1981; to Royston Drift Mine, 23rd March 1982; to South Yorkshire Railway Preservation Society, Attercliffe, Sheffield, 12th August 1987; to SYRPS, Meadowhall, Sheffield, 12th September 1988; displayed at BR Tinsley Depot open day, 29th September 1990; to RMS Locotec, on hire, 14th March 1997, and used at Euro Tunnel; returned to SYRPS, Sheffield, 12th September 1997; to RMS Locotec, Dewsbury, on hire, about January 2001; returned to SYRPS, 2001; to Hanson Quarry Products, Machen, on hire, 5th February 2001; to Peak Rail, Rowsley, 6th April 2006.

D2371   Swindon   1958   52A   12/87 P   D2371
03371   to A.J. Wilkinson, Rowden Mill Station, near Bromyard, Worcestershire, 10th November 1988.

D2373   Swindon   1961   9D   5/68   F   No.1/ DAWN
to NCB Manvers Main Coal Preparation Plant, Wath-on-Dearne, Rotherham (despatched from BR Bolton Depot), September 1968; thereafter worked on lines connecting Barnburgh Main Colliery, Manvers Main Colliery and CPP, and Wath Main Colliery; scrapped on site by E. Nortcliffe Ltd of Rotherham, 1982.

D2381   Swindon   1961   16C   6/72   P   D2381/ 03381
to Flying Scotsman Enterprises, Market Overton, Rutland, 13th April 1973; to Steamtown, Carnforth, 19th March 1976.

D2397   Doncaster   1961   NC   7/87   F   03397
03397   to The Vic Berry Company, Leicester (despatched from BR March), 23rd March 1990; used for spares; scrapped, January 1991.

D2399   Doncaster   1961   NC   7/87   P   03399
03399   to Mangapps Farm Railway Museum, Burnham-on-Crouch, Essex (despatched from BR March), 22nd March 1989.

SECTION 2

Drewry Car Co Ltd 0-6-0 diesel mechanical locomotives built by Vulcan Foundry Ltd, numbered D2200-D2214, and introduced 1952. Fitted with a Gardner 8L3 engine developing 204bhp at 1200rpm, five speed gearbox, and driving wheels of 3ft 3in diameter. Later classified TOPS Class 04.

D2203   DC   2400   1952   ZC   12/67   P   D2203
        VF   D145
to Hemel Hempstead Lightweight Concrete Co Ltd, Cupid Green, Hertfordshire, 2nd February 1968; to Yorkshire Dales Railway, Embsay, 8th February 1982.

D2204   DC   2485   1953   55F   10/69   F   D5
        VF   D211
to Briton Ferry Steel Co Ltd, Glamorgan, March 1970 (sold via W. & F. Smith Ltd, Ecclesfield, Sheffield); scrapped, September 1979.

D2205   DC   2486   1953   51L   7/69   P   D2205/ 11223
        VF   D212
to Tees & Hartlepool Port Authority, Middlesbrough Docks, July 1970; to Kent & East Sussex Railway, 21st August 1983; to West Somerset Railway, Minehead, 18th November 1989; to Somerset & Avon Railway, Radstock, 2nd February 1994; to WSR, Minehead, July 1996.

D2207   DC   2482   1953   ZC   12/67   P   D2207
        VF   D208
to Hemel Hempstead Lightweight Concrete Co Ltd, Cupid Green, Hertfordshire, February 1968; to North Yorkshire Moors Railway, September 1973; to RMS Locotec, Dewsbury, for overhaul, 22nd February 2005; to RMS Locotec, Wakefield, for overhaul, 29th June 2006; to NYMR, 31st January 2007.

D2208   DC   2483   1953   5A   7/68   F   D2208
        VF   D209
to NCB Manvers Coal Preparation Plant, Wath-on-Dearne, Rotherham (despatched from 5B Crewe South Depot), 10th November 1968; to Cortonwood Colliery, Wombwell, March 1969; to Cadeby Colliery, Conisbrough, by May 1969; to Silverwood Colliery, about December 1970; dismantled June 1976; scrapped by a dealer from Worksop, May 1979.

D2209   DC   2484   1953   8J   7/68   F   No.16/ TRACEY
        VF   D210
to NCB Manvers Coal Preparation Plant, Wath-on-Dearne, Rotherham, 10th November 1968; to Kiveton Park Colliery, 13th July 1974; used for spares, and dismantled 1982; scrapped on site by Brinsworth Metals Ltd, 19th August 1985.

D2211   DC   2509   1954   16C   7/70   F   WILF CLEMENT
        VF   D243
to Powell Duffryn Fuels Ltd, NCBOE Coed Bach Disposal Point, August 1970; to Rees Industries Ltd, Llanelli, 3rd August, 1978; scrapped, about November 1980.

D2213   DC   2529   1954   8H   8/68   F   D2213
        VF   D257
to NCB Manvers Coal Preparation Plant, Wath-on-Dearne, Rotherham, September 1969; used as a source of spares; remains scrapped, February 1978.

SECTION 3

Drewry Car Co Ltd 0-6-0 diesel mechanical locomotives built by Vulcan Foundry Ltd and Robert Stephenson & Hawthorns Ltd, numbered D2215-D2273, and introduced 1955. Fitted with a Gardner 8L3 engine developing 204bhp at 1200rpm, five speed gearbox, and driving wheels of 3ft 6in diameter. Later classified TOPS Class 04.

D2219   DC   2542   1955   8H   4/68   F   D2219
        VF   D268
to Barnsley District Coking Co Ltd, Barrow Coking Plant, Barnsley, October 1968; loaned to NCB Barrow Colliery in August 1969, and returned; scrapped, May 1977.

D2225   DC   2548   1955   8F   3/69   F   D2225/ DEBRA
        VF   D274
to NCB Manvers Coal Preparation Plant, Wath-on-Dearne, Rotherham, January 1970; to Wath Colliery, 8th December 1976; scrapped on site by Wath Skip Hire Ltd, July 1985.

D2228   DC   2551   1955   8F   7/68   F   D2228/ 4
        VF   D277
to Bowaters UK Paper Co Ltd, Sittingbourne, Kent, 17th February 1969; scrapped, January to March 1979.

D2229   DC   2552   1955   52A   12/69   P   NPT
        VF   D278
to NCB Brookhouse Colliery, Beighton (despatched from 51L Thornaby Depot), 28th August 1970; to Orgreave Colliery, by 28th July 1971; to Brookhouse Colliery, 12th March 1972; to Orgreave Colliery about October 1973; to Brookhouse Colliery by July 1974; to Manton Colliery, 28th March 1983; to South Yorkshire Railway Preservation Society, Meadowhall, Sheffield, 26th May 1990; to Peak Rail, Rowsley, 12th March 2002.

D2238   DC   2562   1955   8H   7/68   F   D2238/ CAROL
         VF   D288
to NCB Manvers Coal Preparation Plant, Wath-on-Dearne, Rotherham, 10th November 1968; to Coventry Home Fire Plant, Keresley, on loan (as cover while HE 6658 was at Hunslet for repairs), 15th June 1974; returned to Manvers, 5th December 1975; scrapped on site by E. Nortcliffe Ltd of Rotherham, 1982.

D2239   DC   2563   1955   75C   9/71   F   NFT
         VF   D289
to NCB Dodworth Colliery, Barnsley, September 1972; to C.F. Booth Ltd, Rotherham, for scrap, 20th March 1986; scrapped, March 1986.

D2241   DC   2565   1956   30E   5/71   F   2241
         VF   D291
to George Cohen, Sons & Co Ltd, Cransley, Northamptonshire, September 1971; scrapped, November 1976.

D2243   DC   2575   1956   51L   7/69   F   MD2
         RSHN   7862
to Tees & Hartlepool Port Authority, Middlesbrough Docks, July 1970; dismantled 1972; scrapped March 1973.

D2244   DC   2576   1956   55F   6/70   F   5
         RSHN   7863
to A.R. Adams & Son, Newport, Monmouthshire, July 1970; used as a hire loco (see Appendix A); scrapped, January 1981.

D2245   DC   2577   1956   50D   12/68   P   11215
         RSHN   7864
to Derwent Valley Light Railway, Layerthorpe, York, May 1969; to Battlefield Line, Shackerstone, Leicestershire, 17th May 1978.

D2246   DC   2578   1956   55G   7/68   P   D2246
         RSHN   7865
to Coal Mechanisation Ltd, Crawley Depot, Sussex, January 1969; to Tolworth Depot, by September 1990; to British Coal, West Drayton Depot, 19th November 1990; to South Yorkshire Railway Preservation Society (HNRC), Meadowhall, Sheffield, 19th December 1994; to Elsecar Heritage Centre, near Barnsley, on hire, 20th April 1995; returned to SYRPS, 21st August 1996; to South Devon Railway, Buckfastleigh, 9th January 2001.

D2247   DC   2579   1956   55B   11/69   F   D6
         RSHN   7866
to Briton Ferry Steel Co Ltd, Glamorgan, June 1970 (sold via W. & F. Smith Ltd, Ecclesfield, Sheffield); scrapped, September 1979.

D2248   DC   2580   1957   55F   6/70   F   2243/ No.18/ SUE
         RSHN   7867
to NCB Manvers Coal Preparation Plant, Wath-on-Dearne, Rotherham, June 1970; to Maltby Colliery about September 1971; during a repaint at Maltby Colliery received the incorrect number 2243; scrapped by Carol & Good Ltd, Thurcroft, near Rotherham, April 1987.

D2258   DC   2602   1957   16C   9/70   F   D2258/ 4-2
         RSHD   7879
to Hargreaves Industrial Services Ltd, NCBOE Bennerley Disposal Point, Ilkeston, January 1971; to BR Toton Depot, for repairs, December 1974; returned to Bennerley, January 1975; to NCBOE Wentworth Stores, Rotherham, 17th February 1984; to C.F. Booth Ltd, Rotherham, for scrap, 2nd September 1986; scrapped, January 1987.

D2259   DC   2603   1957   73F   12/68   F   D2259/ 5
         RSHD   7889
to Bowaters UK Paper Co Ltd, Sittingbourne, Kent, February 1969; scrapped, January 1978.

D2260   DC   2604   1957   55F   10/70   F   THOMAS HARLING
         RSHD   7890
to Powell Duffryn Fuels Ltd, NCBOE Mill Pit Disposal Point, Cefn Cribbwr, July 1971 (sold via Tilsley & Lovatt Ltd, Trentham); to Cwm Mawr Disposal Point, 3rd November 1981; to Coed Bach Disposal Point, December 1981; scrapped on site by Rees Industries Ltd of Llanelli, June 1983.

D2262   DC   2606   1957   51A   9/68   F   7
         RSHD   7892
to Ford Motor Co Ltd, Dagenham, London, March 1969; involved in a collision; later dismantled for spares and rest scrapped, July 1978.

D2267   DC   2611   1957   50D   12/69   F   No.01
         RSHD   7897
to Ford Motor Co Ltd, Dagenham, London, January 1970; to BR Swindon Works, for rebuild, 19th May 1977; returned to Dagenham, 8th November 1977; dismantled (no engine) by October 1996; to East Anglian Railway Museum, Essex, 24th September 1998; to North Norfolk Railway, Sheringham, 16th February 2000; used for spares; remains scrapped, April 2003.

D2270   DC   2614   1957   55B   2/68   F   D9
         RSHD   7912
to Briton Ferry Steel Co Ltd, Glamorgan, July 1968 (sold via R.E. Trem Ltd, Finningley, Doncaster); scrapped, September 1979.

D2271   DC   2615   1958   55F   10/69   P   D2271
         RSHD   7913
to C.F. Booth Ltd, Rotherham, May 1970; privately purchased for preservation and moved to Thomas Hill Ltd, Kilnhurst, for storage, 27th July 1972; to Midland Railway, Normanton Barracks, Derby, 7th September 1973; to Midland Railway, Butterley, 10th May 1975; to West Somerset Railway, Minehead, 15th May 1982.

D2272   DC   2616   1958   55F   10/70   P   D2272/ ALFIE
         RSHD   7914
to British Fuel Company, Coal Concentration Depot, Blackburn, March 1971; to South Yorkshire Railway Preservation Society (HNRC), Meadowhall, Sheffield, 1st May 1997; to Lavender Line, Isfield, about June 2001; to Peak Rail, Rowsley, by 9th February 2004.

SECTION 4

Drewry Car Co Ltd 0-6-0 diesel mechanical locomotives built by Robert Stephenson & Hawthorns Ltd, numbered D2274-D2340, and introduced 1959. Fitted with a Gardner 8L3 engine developing 204bhp at 1200rpm, five speed gearbox, and driving wheels of 3ft 7in diameter. Later classified TOPS Class 04. Departmental DS1173, built in 1947, was later numbered D2341 and completed Class 04, but some particulars were different from those of the main batch.

D2274	DC	2620	1959	8J	5/69	F	D2274/ No.17
	RSHD	7918					

to NCB Maltby Colliery, near Rotherham, 24th June 1969; scrapped, September 1980.

D2276	DC	2622	1959	30A	8/69	F	D2276
	RSHD	7920					

to A.R. Adams & Son, Newport, Monmouthshire, July 1970; used for spares; remains scrapped, May 1977.

D2279	DC	2656	1960	30E	5/71	P	11249
	RSHD	8097					

to CEGB Rye House Power Station, Hoddesdon, Hertfordshire, 1st October 1971; to East Anglian Railway Museum, Essex, about March 1981.

D2280	DC	2657	1960	30E	3/71	P	D2280
	RSHD	8098					

to Ford Motor Co Ltd, Dagenham, London, 28th June 1971; to BR Swindon Works, for rebuild, 8th July 1977; returned to Dagenham, 8th November 1977; to East Anglian Railway Museum, Essex, 24th September 1998; to North Norfolk Railway, Sheringham, 16th February 2000.

D2281	DC	2658	1960	30E	10/68	F	D2281
	RSHD	8099					

to Briton Ferry Steel Co Ltd, Glamorgan, February 1969 (sold via R.E. Trem Ltd, Finningley, Doncaster); used for spares only; scrapped, August 1971.

D2284	DC	2661	1960	30E	4/71	P	D2284
	RSHD	8102					

to NCB North Gawber Colliery, Mapplewell, Barnsley, 16th July 1971; to Grimethorpe Colliery, 30th January 1976; to Woolley Colliery, March 1978; to South Yorkshire Railway Preservation Society, Chapeltown, 2nd August 1985; to SYRPS, Attercliffe, Sheffield, December 1986; to SYRPS, Meadowhall, Sheffield, 12th September 1988; to Peak Rail, Rowsley, March 2002.

D2294	DC	2674	1960	70D	2/71	F	01
	RSHD	8127					

to Shipbreakers (Queenborough) Ltd, Kent, 16th March 1972; scrapped, October 1985.

D2298	DC	2679	1960	52A	12/68	P	D2298
	RSHD	8157					

to Derwent Valley Light Railway, Layerthorpe, York, April 1969; to Quainton Railway Society, Buckinghamshire, 22nd October 1982.

D2299 DC   2680   1960   52A   1/70   F D2299/ JONAH
RSHD   8158
to NCB Bestwood Colliery, Nottinghamshire (despatched from BR Thornaby Depot), 9th July 1970; to Hucknall Colliery, 7th August 1970; to Calverton Colliery, May 1978; to Hucknall Colliery, 23rd August 1978; to C.F. Booth Ltd, Rotherham, for scrap, February 1984; scrapped, week ending 17th February 1984.

D2300 DC   2681   1960   8J   5/69   F D2300
RSHD   8159
to NCB Shireoaks Colliery, 25th June 1969; to Steetley Colliery, on loan, 12th September 1974; returned to Shireoaks Colliery, 18th November 1974; to Manton Colliery, 18th October 1978; scrapped by Hoyland Dismantling Co Ltd, August 1986.

D2302 DC   2683   1960   16C   6/69   P D2302
RSHD   8161
to British Sugar Corporation Ltd, Woodston Factory, Peterborough, August 1969; to BSC Allscott Factory, Shropshire, October 1969; to G.G. Papworth Ltd, Ely, Cambridgeshire, 12th July 1983; to South Yorkshire Railway Preservation Society (HNRC), Meadowhall, Sheffield, 25th September 1993; to Rutland Railway, on loan, 16th March 2001; to Barrow Hill Engine Shed Society, Staveley, 18th May 2004.

D2304 DC   2685   1960   51A   2/68   F D8
RSHD   8163
to Llanelly Steel Co Ltd, Carmarthenshire, July 1968 (sold via R.E. Trem Ltd, Finningley, Doncaster); scrapped, May 1977.

D2305 DC   2686   1960   51A   2/68   F D9
RSHD   8164
to Llanelly Steel Co Ltd, Carmarthenshire, May 1968 (sold via R.E. Trem Ltd, Finningley, Doncaster); scrapped, about September 1981.

D2306 DC   2687   1960   51L   2/68   F D6
RSHD   8165
to Llanelly Steel Co Ltd, Carmarthenshire, 11th July 1968 (sold via R.E. Trem Ltd, Finningley, Doncaster); scrapped, about September 1981.

D2307 DC   2688   1960   51L   2/68   F D7
RSHD   8166
to Llanelly Steel Co Ltd, Carmarthenshire, July 1968 (sold via R.E. Trem Ltd, Finningley, Doncaster); scrapped, October 1979.

D2308 DC   2689   1960   51A   2/68   F D8
RSHD   8167
to Briton Ferry Steel Co Ltd, Glamorgan, August 1968 (sold via R.E. Trem Ltd, Finningley, Doncaster, and passed through Rotherham on a low-loader, 24th August 1968); to Duport Steel Works Ltd, Llanelli, 25th October 1979; scrapped, May 1980.

D2310   DC   2691   1960   52A   1/69   P   D2310/ 04110
        RSHD   8169
to Coal Mechanisation Ltd, Tolworth, London, April to June 1969; to South Yorkshire Railway Preservation Society (HNRC), Meadowhall, Sheffield, 14th September 1994; to Battlefield Line, Shackerstone, 3rd October 2001.

D2317   DC   2698   1960   52A   8/69   F   No.10
        RSHD   8176
to NCB Manvers Main Coal Preparation Plant, Wath-on-Dearne, Rotherham, 30th December 1969; to Cortonwood Colliery, Wombwell, about 5th May 1970; scrapped on site by Wath Skip Hire Ltd, July 1986.

D2322   DC   2703   1961   52A   8/68   F   D2322/ No.24
        RSHD   8181
to NCB Orgreave Colliery, February 1969; to Treeton Colliery, 1972; to Orgreave Colliery, about January 1973; to Treeton Colliery, about October 1973; to Orgreave Colliery, by April 1974; to New Stubbin Colliery, on loan, summer 1975; returned to Orgreave; to Kiveton Park Colliery, 29th April 1980; scrapped, by 28th November 1985.

D2324   DC   2705   1961   55B   7/68   P   2324/JUDITH
        RSHD   8183
to G.W. Talbot Ltd, Coal Concentration Depot, Aylesbury, January 1969; to Redland Roadstone Ltd, Barrow upon Soar, Leicestershire, about November 1989; to South Yorkshire Railway Preservation Society (HNRC), Meadowhall, Sheffield, 29th March 1995; to Lavender Line, Isfield, about June 2001; to Barrow Hill Engine Shed Society, Staveley, 27th March 2006; to Peak Rail, Rowsley, 1st October 2008.

D2325   DC   2706   1961   50D   7/68   P   D2325
        RSHD   8184
to NCB Norwich Coal Concentration Depot, December 1968; to Tannick Commercial Repairs, Norwich, for storage, November 1986; to John Jolly, Bridgewick Farm, Dengie, Southminster, Essex, 19th March 1987; to Mangapps Farm Railway Museum, Burnham-on-Crouch, Essex, 1988.

D2326   DC   2707   1961   52A   8/68   F   D2326
        RSHD   8185
to NCB Manvers Main Coal Preparation Plant, Wath-on-Dearne, Rotherham, February 1969; used for spares; remains scrapped, Autumn 1975.

D2327   DC   2708   1961   52A   8/68   F   No.12/ 521-12
        RSHD   8186
to NCB Manton Main Colliery, 1969; to Dinnington Colliery, 9th August 1971; to NCB Elsecar Workshops, 3rd May 1973; returned to Dinnington Colliery; to Elsecar Workshops, 15th November 1974; to Dinnington Colliery, 20th January 1975; to Coopers (Metals) Ltd, Sheffield, for scrap, 5th January 1984; scrapped, February 1984.

D2328　DC　　　　2709　1961　52A　9/68　F　No.31
RSHD　　8187
to NCB Dinnington Colliery, 6th June 1969; to Steetley Colliery, April 1973; to BR Doncaster Works for wheel turning, 2nd February 1977; returned to Steetley Colliery, 10th February 1977; to Shireoaks Colliery by April 1982; to Kiveton Park Colliery, 13th May 1982; to Cortonwood Colliery, 18th July 1985; scrapped on site by Wath Skip Hire Ltd, 1986.

D2329　DC　　　　2710　1961　52A　7/68　F　D2329
RSHD　　8188
to Derwent Valley Light Railway, Layerthorpe, York, January 1969 (sold via Peter Wood & Co Ltd, Eckington, Sheffield); used for spares; remains scrapped, April 1970.

D2332　DC　　　　2713　1961　52A　6/69　F　D2332/ LLOYD
RSHD　　8191
to NCB Manvers Main Coal Preparation Plant, Wath-on-Dearne, Rotherham, January 1970; to Cadeby Colliery, 28th August 1975; to Thurcroft Colliery, 14th June 1976; to Shireoaks Colliery, 29th June 1981; to Thurcroft Colliery, 3rd September 1982; to Dinnington Colliery, 19th July 1985; scrapped, July 1986.

D2333　DC　　　　2714　1961　52A　9/69　F　3/ P1062C
RSHD　　8192
to Ford Motor Co Ltd, Dagenham, London, December 1969; to BR Swindon Works, for rebuild, 3rd May 1977; returned to Dagenham, 23rd January 1978; scrapped, early 1990.

D2334　DC　　　　2715　1961　51A　7/68　P　D2334
RSHD　　8193
to NCB Manvers Main Coal Preparation Plant, Wath-on-Dearne, Rotherham, 2nd June 1969; to Thurcroft Colliery, 8th October 1969; to Dinnington Colliery, 19th July 1985; to Maltby Colliery, 24th February 1986; to South Yorkshire Railway Preservation Society, Meadowhall, Sheffield, 12th November 1988; to Knights of Old Ltd, Old, Northamptonshire, 28th September 1993; to Churnet Valley Railway, Cheddleton, 10th July 1994.

D2335　DC　　　　2716　1961　51A　7/68　F　2
RSHD　　8194
to NCB Manvers Main Coal Preparation Plant, Wath-on-Dearne, Rotherham, 2nd June 1969; to Maltby Colliery, about September 1969; scrapped, about September 1980.

D2336　DC　　　　2717　1961　51A　7/68　F　D2336
RSHD　　8195
to NCB Manvers Main Coal Preparation Plant, Wath-on-Dearne, Rotherham, 2nd June 1969; used for spares; remains scrapped, about February 1978.

D2337　DC　　　　2718　1961　51A　7/68　P　D2337
RSHD　　8196
to NCB Manvers Main Coal Preparation Plant, Wath-on-Dearne, Rotherham, 2nd June 1969; to Barnburgh Main Colliery, June 1974; to Manvers Main, February 1977; to South Yorkshire Railway Preservation Society, Attercliffe, Sheffield, 22nd February 1988; to SYRPS, Meadowhall, Sheffield, 15th September 1988; to Peak Rail, Rowsley, 15th March 2002.

D2340   DC   2593   1956   55F   10/68 F   D1
        RSHD   7870
Demonstration locomotive; sold to British Railways, March 1962; to Briton Ferry Steel Co Ltd, Glamorgan, April 1969; scrapped, September 1979.

SECTION 5

Andrew Barclay, Sons & Co Ltd built 0-4-0 diesel mechanical locomotives, numbered D2410-D2444, and introduced 1958. Fitted with a Gardner 8L3 engine developing 204bhp at 1200rpm; five speed gearbox, and driving wheels of 3ft 7in diameter. Later classified TOPS Class 06.

D2420   AB   435   1959   RSD   1984   P   06003
06003   withdrawn as BR Departmental loco 97804 at Reading Signal Depot; to C.F. Booth Ltd, Rotherham (despatched from 81A Old Oak Common Depot), 25th September 1986; to South Yorkshire Railway Preservation Society (HNRC), Attercliffe, Sheffield, 9th March 1987; to SYRPS, Meadowhall, Sheffield, 14th September 1988; displayed at BR Tinsley Depot open day, 29th September 1990; returned to SYRPS; to Crewe Works open day, 2nd May 1997; to Battlefield Line, Shackerstone, on loan, 9th May 1997; returned to SYRPS, 23rd October 1997; to Barrow Hill Engine Shed Society, Staveley, 11th June 1999; to Rutland Railway Museum, Cottesmore, 9th September 1999; to Barrow Hill Engine Shed Society, Staveley, 13th September 2002; to UK Coal, Widdrington Disposal Point, on hire, 18th July 2003; to Barrow Hill Engine Shed Society, Staveley, 28th June 2006; to Peak Rail, Rowsley, May 2008.

D2432   AB   459   1960   65A   12/68 F   NFT
to Shipbreakers (Queenborough) Ltd, Kent, 20th May 1969; exported from Sheerness Docks to Italy, March 1977. (see Appendix C).

SECTION 6

Hudswell, Clarke & Co Ltd built 0-6-0 diesel mechanical locomotives, numbered D2510-D2519, and introduced 1961. Fitted with a Gardner 8L3 engine developing 204bhp at 1200rpm; four speed gearbox, and driving wheels of 3ft 6in diameter. No TOPS classification.

D2511   HC   D1202   1961   12C   12/67 P   D2511
to NCB Brodsworth Colliery, Doncaster, May 1968; to Keighley & Worth Valley Railway, Haworth, West Yorkshire, 8th October 1977.

D2513   HC   D1204   1961   12C   8/67   F   D2513
to NCB Cadeby Colliery, Conisbrough, December 1968; scrapped, October 1975.

D2518   HC   D1209   1962   5A   2/67   F   D2518
to NCB Hatfield Colliery, Doncaster, August 1967; later used for spares; scrapped, June 1973.

D2519   HC   D1210   1962   5A   7/67   F   D2519
to NCB Hatfield Colliery, Doncaster, February 1968; to Keighley & Worth Valley Railway, Haworth, West Yorkshire, 3rd April 1982; to Marple & Gillott Ltd, Sheffield, for scrap, 27th March 1985; scrapped, April 1985.

SECTION 7

Hunslet Engine Co Ltd built 0-6-0 diesel mechanical locomotives, numbered D2550-D2618, and introduced 1955. Fitted with a Gardner 8L3 engine developing 204bhp at 1200rpm, four speed gearbox, and driving wheels of 3ft 4in diameter (D2550-D2573) and 3ft 9in (D2574-D2618). Later classified TOPS Class 05.

D2554   HE   4870   1956   70H   9/83   P   D2554
05001   withdrawn by BR as Departmental locomotive No.97803 on Isle of Wight; to Isle of Wight Steam Railway, Havenstreet, 27th August 1984.

D2561   HE   4999   1957   8F   8/67   F   D3
to Llanelly Steel Co Ltd, Carmarthenshire, March 1968; scrapped, October 1972.

D2568   HE   5006   1957   8F   8/67   F   D2568
to Briton Ferry Steel Co Ltd, Glamorgan, 9th May 1968; scrapped, about May 1969.

D2569   HE   5007   1957   8F   8/67   F   D6
to Briton Ferry Steel Co Ltd, Glamorgan (despatched from 8C Speke Junction Depot), May 1968; scrapped, about May 1969.

D2570   HE   5008   1957   8F   7/67   F   NFT
to Briton Ferry Steel Co Ltd, Glamorgan (despatched from 8C Speke Junction Depot), March 1968; scrapped, June 1971.

D2578   HE   5460   1958   62A   7/67   P   D2578
to Hunslet Engine Co Ltd, Leeds, December 1967; rebuilt as Hunslet 6999; to H.P. Bulmer Ltd, Cider Manufacturers, Hereford, July 1968; to private site, Croes Newydd, Wrexham, November 2000; to D2578 Locomotive Group, Moreton on Lugg, 6th August 2001.

D2587   HE   5636   1959   62C   12/67   P   D2587
to Hunslet Engine Co Ltd, Leeds, September 1968; rebuilt (Hunslet 7180 of 1969) with a 384hp engine; to CEGB Chadderton Power Station, September 1969; to Kearsley Power Station, 3rd November 1981; to East Lancashire Railway, Bury, March 1983; to South Yorkshire Railway Preservation Society (HNRC), Meadowhall, Sheffield, 30th August 1997; to Lavender Line, Isfield, about June 2001; to Peak Rail, Rowsley, by 9th February 2004.

D2595   HE   5644   1960   62A   6/68   P   D2595
to Hunslet Engine Co Ltd, Leeds, January 1969; rebuilt (Hunslet 7179 of 1969) with a 384hp engine; to CEGB Chadderton Power Station, September 1969; to Kearsley Power Station, 24th September 1981; to East Lancashire Railway, Bury, March 1983; to Steamport Transport Museum, Southport, Merseyside, 19th October 1989; to Ribble Steam Railway, Preston, 31st March 1999.

D2598   HE   5647   1960   50D   12/67   F   SAM
to NCB Rossington Colliery, Doncaster, 17th May 1968; to Askern Colliery, July 1971; to Lambton Engine Works, Tyne & Wear, February 1975; scrapped, May 1975.

D2599   HE   5648   1960   50D   12/67   F   F/SE/357
to NCB Hickleton Colliery, Doncaster, May 1968; to Frickley Colliery about October 1968; to Askern Colliery, 16th June 1976; scrapped on site by Geeson Ltd of Ripley, May 1981.

D2600   HE   5649   1960   50D   12/67   F   D7
to Briton Ferry Steel Co Ltd, Glamorgan (sold via R.E. Trem Ltd, Finningley, Doncaster and left 50D Goole Depot on 29th April 1968); scrapped, June 1971.

D2601   HE   5650   1960   50D   12/67   F   D5
to Llanelly Steel Co Ltd, Carmarthenshire (sold via R.E. Trem Ltd, Finningley, Doncaster and left 50D Goole Depot on 29th April 1968); arrived at Llanelli, 3rd May 1968; scrapped, 1979.

D2607   HE   5656   1960   6G   12/67   F   D2607
to NCB Dinnington Colliery, 9th September 1968; to Steetley Colliery about October 1968; to Shireoaks Colliery, on loan, May 1971; returned to Steetley Colliery, August 1971; to Fence Workshops for overhaul, 28th May 1974; returned to Steetley Colliery, 30th October 1974; to Shireoaks Colliery, on loan, 17th May 1975; returned to Steetley Colliery, July 1975; to Treeton Colliery, on loan, 3rd December 1975; returned to Steetley Colliery, December 1975; to BR Doncaster, for wheel turning, August 1977; returned to Steetley Colliery about September 1977; to Shireoaks Colliery, on loan, 26th August 1980; returned to Steetley Colliery, 16th January 1981; to Treeton Colliery, on loan, April 1981; returned to Steetley Colliery, July 1981; to Coopers (Metals) Ltd, Sheffield, for scrap, 12th June 1984; scrapped, by 4th July 1984.

D2611   HE   5660   1960   50D   12/67   F   D2611
to NCB Yorkshire Main Colliery, Doncaster, May 1968; scrapped, about December 1976.

D2613   HE   5662   1960   50D   12/67   F   D2613/ BRM5481
to NCB Brodsworth Colliery, Doncaster, May 1968; to Bentley Colliery, 1974; scrapped on site by W. Heselwood Ltd of Sheffield, June 1977.

D2616   HE   5665   1961   50D   12/67   F   D2616
to NCB Hatfield Colliery, Doncaster, May 1968; extant April 1973; scrapped later in 1973.

D2617   HE   5666   1961   62C   12/67   F   D2617
to Hunslet Engine Co Ltd, Leeds, autumn 1968; used for spares; remains scrapped, April 1976.

SECTION 8

North British Locomotive Co Ltd built 0-4-0 diesel hydraulic locomotives, numbered D2708-D2780, and introduced 1957. Fitted with a North British/M.A.N. W6V 17.5/22A engine developing 225bhp at 1100rpm, and driving wheels of 3ft 6in diameter. No TOPS classification.

D2720   NB   27815   1958   64H   7/67   F   NFT
to James N. Connell Ltd, Coatbridge, Lanarkshire, June 1968; scrapped, 1971.

D2726   NB   27821   1958   ZN   2/67   F   NFT
to Shipbreakers (Queenborough) Ltd, Kent, October 1967 (sold via R.E. Trem Ltd, Finningley, Doncaster); scrapped, October 1971.

D2736   NB   27831   1958   65A   3/67   F   D2736
to Bird's Commercial Motors Ltd, Long Marston, Worcestershire, July 1967; to Bird's (Swansea) Ltd, Pontymister Works, Risca, August 1967; to Bird's, 40 Acre Site, Cardiff, 25th February 1968; scrapped, July 1969.

D2738   NB   27833   1958   65A   6/67   F   NFT
to Andrew Barclay, Sons & Co Ltd, Kilmarnock, October 1967; rebuilt 1968; to NCB Killoch Colliery, Ochiltree, about July 1969; sold to Alex Smith Metals of Ayr, by March 1979; scrapped on site, January 1980.

D2739   NB   27834   1958   65A   3/67   F   D2739
to Bird's Commercial Motors Ltd, Long Marston, Worcestershire, July 1967; scrapped, September 1969.

D2757   NB   28010   1960   65A   7/67   F   NFT
to Bird's (Swansea) Ltd, Pontymister Works, Risca, 9th November 1967; to Bird's, 40 Acre Site, Cardiff, February 1968; scrapped, October 1970.

D2763   NB   28016   1960   65A   6/67   F   NFT
to Andrew Barclay, Sons & Co Ltd, Kilmarnock, October 1967; rebuilt 1968; to BSC Landore Foundry, Swansea, by September 1969; scrapped, April 1977.

D2767   NB   28020   1960   65A   6/67   P   D2767
to Andrew Barclay, Sons & Co Ltd, Kilmarnock, October 1967; rebuilt 1968; to Burmah Oil Co Ltd, Stanlow, Cheshire, 24th April 1969; to East Lancashire Railway, Bury, 12th June 1983; to Manchester Metrolink, on hire, October 1991; returned to ELR, 1991; to Scottish Railway Preservation Society, Bo'ness, 25th July 2001.

D2774   NB   28027   1960   65A   6/67   P   D2774
to Andrew Barclay, Sons & Co Ltd, Kilmarnock, October 1967; rebuilt 1968; to NCB Killoch Colliery, on hire, about July 1969; returned to Andrew Barclay, 1971; sold to NCB; to Celynen North Colliery, Newbridge, early April 1971; to BR Canton Depot, Cardiff, for repairs, March 1976; noted at Hafodyrynys Colliery on 26th March 1976; to Celynen North Colliery, April 1976; to Celynen South Colliery, Abercarn, 4th September 1976; to BR Canton Depot, Cardiff, for wheel turning, 17th May 1982; to NCB Mountain Ash Works, 27th May 1982; to Celynen South Colliery, March 1983; to East Lancashire Railway, Bury, 4th October 1986; to Strathspey Railway, 2nd May 2001.

D2777   NB   28030   1960   65A   3/67   F   D2777
to Bird's Commercial Motors Ltd, Long Marston, Worcestershire, July 1967; to Bird's (Swansea) Ltd, Pontymister Works, Risca, November 1967; scrapped, 1968.

SECTION 9

Yorkshire Engine Co Ltd built 0-4-0 diesel hydraulic locomotives, numbered D2850-D2869, and introduced 1960. Fitted with a Rolls-Royce C6NFL engine developing 179bhp at 1800rpm, and driving wheels of 3ft 6in diameter. Later classified TOPS Class 02.

D2853   YE   2812   1960   8J   6/75   P   02003/ PETER
02003   to L.C.P. Fuels Ltd, Shut End Works, West Midlands, 19th November 1975;

displayed at BR Bescot open day, 9th October 1988; returned to LCP; to South Yorkshire Railway Preservation Society (HNRC), Meadowhall, Sheffield, 18th April 1997; to Rutland Railway, 22nd June 2001; to Barrow Hill Engine Shed Society, Staveley, 1st September 2003; to Appleby Frodingham Railway Society, Scunthorpe, on loan, 24th March 2006.

D2854   YE   2813   1960   8J   2/70   P   D2854
to C.F. Booth Ltd, Rotherham, 29th August 1970; to South Yorkshire Railway Preservation Society (HNRC), Attercliffe, Sheffield, May 1988; to SYRPS, Meadowhall, Sheffield, 14th September 1988; to Middleton Railway, Leeds, on loan, 10th September 1994; returned to SYRPS, 15th October 1994; to Sheffield Supertram, Nunnery Depot, on hire, 7th November 1994; returned to SYRPS, 27th May 1995; to Peak Rail, Darley Dale, 2001; to Peak Rail, Rowsley, March 2002.

D2856   YE   2815   1960   8J   6/75   F   02004
02004   to Redland Roadstone Ltd, Mountsorrel, Leicestershire, 25th September 1975; used for spares; remains (minus engine) to Budden Wood Quarry, Leicestershire; scrapped on site by The Vic Berry Company of Leicester, 1986.

D2857   YE   2816   1960   8J   4/71   F   NFT
to Bird's Commercial Motors Ltd, Long Marston, Worcestershire, November 1971; extant, 31st March 1992; scrapped, later in 1992.

D2858   YE   2817   1960   9A   2/70   P   D2858
to Hutchinson Estate & Dock Co (Widnes) Ltd, Widnes (despatched from 9D Newton Heath), August 1970; to Fisons Fertilisers, Widnes, September 1978; to Lowton Metals Ltd, Haydock, 5th March 1981; to Butterley Engineering Ltd, Ripley, Derbyshire, November 1986; to Midland Railway, Butterley, 14th June 2002.

D2860   YE   2843   1961   8J   12/70   P   D2860
to Curator of Historical Relics, BR Preston Park, Brighton, March 1973; to National Railway Museum, York, 26th November 1977; to Thomas Hill (Rotherham) Ltd, Kilnhurst, for overhaul and repaint, 14th September 1978; to National Railway Museum, York, 3rd January 1979 to Gloucestershire Warwickshire Railway, for gala, 9th July 2009; returned to NRM, York, mid July 2009.

D2862   YE   2845   1961   10D   12/69   F   ND3/ 63000359
to Tilsley & Lovatt Ltd, Trentham, Staffordshire, March 1970; overhauled and sold to NCB; to Norton Colliery, Staffordshire, January 1971; to Chatterley Whitfield Colliery, Tunstall, week ending 25th September 1971; to Norton Colliery, week ending 15th October 1971; scrapped, about April 1979.

D2865   YE   2848   1961   50D   3/70   F   NFT
to APCM, Kilvington, Nottinghamshire, September 1970; to Blue Circle, Beeston Depot, Nottingham, for storage, 1984; to The Vic Berry Company, Leicester, for scrap, December 1984; scrapped, May 1985.

D2866   YE   2849   1961   9A   2/70   P   NPT
to Arnott Young Ltd, Dalmuir, Dunbartonshire (despatched from 9D Newton Heath), about August 1970; to BR Glasgow Works, for repairs, April 1977; returned to Dalmuir; to Caledonian Railway Ltd, Brechin, Tayside, 17th October 1987; to South Yorkshire Railway Preservation Society (HNRC), Meadowhall, Sheffield, 22nd January 1996; to Peak Rail, Rowsley, March 2002.

D2867 YE 2850 1961 6A 9/70 P DIANE
to Tunnel Cement, Pennyford, on hire, August 1970; to Redland Roadstone Ltd, Mountsorrel, Leicestershire, 23rd October 1970; to Barrow-upon-Soar Works, Leicestershire, late 1979; to South Yorkshire Railway Preservation Society (HNRC), Meadowhall, Sheffield, 31st March 1995; to Battlefield Line, Shackerstone, 27th July 2001.

D2868 YE 2851 1961 10D 12/69 P D2868
to Lunt, Comley and Pitt Ltd, Shut End Works, Staffordshire, October 1970; to South Yorkshire Railway Preservation Society (HNRC), Meadowhall, Sheffield, 25th April 1997; to Lavender Line, Isfield, about June 2001; to Barrow Hill Engine Shed Society, Staveley, 28th January 2004; to Peak Rail, Rowsley, 13th May 2008.

SECTION 10

Hunslet Engine Co Ltd built 0-4-0 diesel mechanical locomotives, numbered D2950-D2952, and introduced 1954. Fitted with a Gardner 6L3 engine developing 153bhp at 1200rpm, four speed gearbox, and driving wheels of 3ft 4in diameter. No TOPS classification.

D2950 HE 4625 1954 50D 12/67 F D4
to Llanelly Steel Co Ltd, Carmarthenshire (sold via R.E. Trem Ltd, Finningley, Doncaster and left Goole shed 29th April 1968); arrived at Llanelli, 2nd May 1968; stored at Thyssen Ltd, Old Castle Depot, Llanelli, May 1980; scrapped by Gwillym Jones & Son, spring 1983; engine used in a trawler.

SECTION 11

Andrew Barclay, Sons & Co Ltd built 0-4-0 diesel mechanical locomotives, numbered D2953-D2956, and introduced 1956. Fitted with a Gardner 6L3 engine developing 153bhp at 1200rpm, four speed gearbox, and driving wheels of 3ft 2in diameter. BR Departmental locomotive number 81 became the second D2956 after the first one was withdrawn. Later classified TOPS Class 01.

D2953 AB 395 1955 30A 6/66 P D2953
to Thames Matex Ltd, West Thurrock, Essex, June 1966 (the first BR diesel shunter sold for industrial service); to BP Refinery (Kent) Ltd, Grain, on loan, 1967; returned to Thames Matex; to Shell Mex & BP Ltd, Purfleet, on loan, at various times; to South Yorkshire Railway Preservation Society, Chapeltown, 15th December 1985; to SYRPS, Attercliffe, Sheffield, December 1986; to SYRPS, Meadowhall, Sheffield, 15th September 1988; to Peak Rail, Rowsley, March 2002.

D2956 AB 398 1956 36A 5/66 P D2956
to A. King & Sons Ltd, Norwich, July 1966; to A. King & Sons Ltd, Snailwell, Cambridgeshire, September 1981; to East Lancashire Railway, Bury, 30th July 1985.

D2956 AB 424 1958 36A 11/67 F D5
the second D2956, withdrawn by BR as Departmental locomotive number 81; to Briton Ferry Steel Co Ltd, Glamorgan, March 1968; scrapped, August 1969.

31

SECTION 12

Ruston & Hornsby Ltd built 0-4-0 diesel mechanical locomotives (Ruston's class 165DS), numbered D2957-D2958, and introduced 1956. Fitted with a Ruston 6VPHL engine developing 165bhp at 1250rpm, four speed gearbox, and driving wheels of 3ft 4in diameter. No TOPS classification.

D2958   RH   390777 1956   30A   1/68   F   NFT
to C.F. Booth Ltd, Rotherham, May 1968 (sold via R.E. Trem Ltd, Finningley, Doncaster); scrapped, October 1984.

SECTION 13

Ruston & Hornsby Ltd built 0-6-0 diesel electric locomotives (Ruston's class LSSE), numbered D2985-D2998, and introduced 1962. Fitted with a Paxman 6RPHL engine developing 275bhp at 1360rpm, and driving wheels of 3ft 6in diameter. Later classified TOPS Class 07.

D2985   RH   480686 1962   70D   7/77   P   NPT
07001   to Tilsley & Lovatt Ltd, Trentham, Staffordshire, for overhaul, about April 1978; to Peakstone Ltd, Holderness Limeworks, Peak Dale, Derbyshire, 30th May 1978; to South Yorkshire Railway Preservation Society (HNRC), Meadowhall, Sheffield, June 1989; to Mayer Parry Ltd, Snailwell, on hire, 28th April 1993; returned to SYRPS, 22nd October 1997; to Barrow Hill Engine Shed Society, Staveley, 30th June 1999; to Creative Logistics, Salford, on hire, 5th March 2001; to Barrow Hill Engine Shed Society, Staveley, 20th July 2009.

D2986   RH   480687 1962   70D   7/77   F   NFT
07002   to Powell Duffryn Fuels Ltd, NCBOE Coed Bach Disposal Point, Kidwelly, Dyfed, April 1978; scrapped on site by T. Davies of Llanelli, September 1982.

D2987   RH   480688 1962   70D   10/76   F   07003
07003   to R.E. Trem Ltd, Finningley, Doncaster, March 1977; to British Industrial Sand Ltd, Oakamoor, Staffordshire, about October 1978; scrapped, May 1985.

D2989   RH   480690 1962   70D   7/77   P   LANGBAURGH
07005   to Resco (Railways) Ltd, Woolwich, for overhaul (works number L106 of 1978), June 1978; to ICI Wilton Works, Middlesbrough, 18th July 1979; to Barrow Hill Engine Shed Society (HNRC), Staveley, 21st December 2000; to Battlefield Line, Shackerstone, 3rd September 2003; to Great Central Railway, Loughborough, 21st May 2008.

D2990   RH   480691 1962   70D   7/77   F   NFT
07006   to Powell Duffryn Fuels Ltd, NCBOE Coed Bach Disposal Point, Kidwelly, Dyfed, April 1978; scrapped on site by T. Davis of Llanelli, October 1984.

D2991   RH   480692 1962   70D   5/73   P   07007
07007   to Eastleigh Works, 1973; used as a stationary generator, then placed in store; to Eastleigh Railway Preservation Society, Eastleigh Works, about September 1988; to Knights Rail Services, Eastleigh Works, about March 2007; major overhaul, January 2008; resumed working, 19th February 2008; to Swanage Railway, for gala, 6th May 2008; returned to Eastleigh, 16th May 2008.

D2994   RH   480695 1962   70F   10/76   P   D2994
07010   to Resco (Railways) Ltd, Woolwich, for overhaul (despatched from BR Eastleigh), June 1978; to Winchester & Alton Railway, New Alresford, Hampshire, August 1978; to West Somerset Railway, 19th May 1980; to Avon Valley Railway, Bitton, Gloucestershire, 1st March 1994.

D2995   RH   480696 1962   70D   7/77   P   07011/CLEVELAND
07011   to Resco (Railways) Ltd, Woolwich, for overhaul (works number L105 of 1978), June 1978; to ICI Billingham Works, on hire, March 1979; returned to Resco, 12th November 1979; to ICI Wilton Works, Middlesbrough, 4th September 1980; to Hastings Diesels, St Leonards, East Sussex, 17th May 1996; to Kent & East Sussex Railway, Tenterden, Kent, 1998; to St Leonards Railway Engineering Ltd, about July 2000.

D2996   RH   480697 1962   70D   7/77   P   07012
07012   to Powell Duffryn Fuels Ltd, NCBOE Cwm Mawr Disposal Point, Tumble, Dyfed, April 1978; to NCBOE Coed Bach Disposal Point, early 1982; to South Yorkshire Railway Preservation Society (HNRC), Meadowhall, Sheffield, 11th December 1992; to Barrow Hill Engine Shed Society, 28th June 1999: to Lavender Line, Isfield, for storage, about June 2001; to Barrow Hill Engine Shed Society, Staveley, 18th July 2006; to Appleby Frodingham Railway Society, Scunthorpe, about January 2009.

D2997   RH   480698 1962   70D   7/77   P   07013
07013   to Resco (Railways) Ltd, Woolwich, for overhaul (works number L101 of 1978), May 1978; to Dow Chemical Co Ltd, King's Lynn, 5th October 1978; to South Yorkshire Railway Preservation Society (HNRC), Meadowhall, Sheffield, 16th August 1994; to Barrow Hill Engine Shed Society, Staveley, 28th June 1999; to Peak Rail, Rowsley, 24th October 2003.

SECTION 14

British Railways built 0-6-0 diesel electric locomotives, numbered D3000-D3116, and introduced 1953. Fitted with an English Electric 6KT engine developing 350bhp at 630rpm, and driving wheels of 4ft 6in diameter. Later classified TOPS Class 08.

D3000   Derby   1952   82A   11/72   P   D3000
to NCB Hafodyrynys Colliery, Pontypool, 19th March 1973; to BR Canton Depot, Cardiff, for repairs, 11th August 1975; to Hafodyrynys Colliery, about November 1975; to Bargoed Colliery, Mid Glamorgan, 17th July 1978; to BR Canton Depot, Cardiff, October 1979; to Bargoed Colliery, October 1979; to BR Canton Depot, Cardiff, June 1980; to Bargoed Colliery, July 1980; to Mountain Ash Colliery, 10th July 1981; to Mardy Colliery, 13th November 1981; to Mountain Ash Colliery, December 1981; to Mardy Colliery, May 1982; to Brighton Railway Museum, despatched 18th March 1987, arrived 21st March 1987; to South Yorkshire Railway Preservation Society, Meadowhall, Sheffield, 9th March 1993; to Barrow Hill Engine Shed Society, Staveley, 7th March 2001; to Appleby Frodingham RPS, Scunthorpe, 14th April 2008; to Peak Rail, Rowsley, 27th January 2011.

D3002   Derby   1952   82A   7/72   P   13002
to Foster Yeoman Quarries Ltd, Merehead Stone Terminal, Somerset, November 1972;

at BR Westbury Depot, February 1976; to BR Bath Road Depot, Bristol, for repairs, 13th April 1976; returned to Merehead, June 1976; to Plym Valley Railway Association, Marsh Mills, Plympton, Devon, 9th July 1982.

D3003   Derby   1952   82A   7/72   F   MEREHEAD
to Foster Yeoman Quarries Ltd, Merehead Stone Terminal, Somerset, May 1973; to BR Derby Works, for repairs, 30th June 1974; returned to Merehead, late 1974; to Wanstrow Childrens Playground, about February 1982; scrapped, December 1991.

D3011   Derby   1952   70D   10/72   F   LICKEY
to British Leyland Ltd, Longbridge, Birmingham, 8th January 1973; to BR Derby Works, for repair, 15th January 1976; returned April 1976; to BR Tyseley Depot, for wheel turning, November 1981; returned to Longbridge; to Marple & Gillott Ltd, Sheffield, 6th December 1985; scrapped, December 1985.

D3014   Derby   1952   70D   10/72   P   D3014/SAMSON
to NCB Merthyr Vale Colliery, Aberfan, August to September 1973; to BR Canton Depot, Cardiff, for repairs, December 1974; to Merthyr Vale Colliery, about January 1975; to BR Canton Depot, Cardiff, for repairs, 19th February 1980; to Merthyr Vale Colliery, 21st May 1980; to BR Canton Depot, Cardiff, for repairs, September 1980; to Merthyr Vale Colliery, October 1980; to BR Canton Depot, Cardiff, for repairs, 3rd October 1981; to Merthyr Vale Colliery, 27th December 1981; to BR Canton Depot, Cardiff, for repairs, 19th July 1985; to Merthyr Vale Colliery, 28th March 1986; to BR Canton Depot, Cardiff, for repairs, 5th October 1987; to Merthyr Vale Colliery, 30th October 1987; to Paignton & Dartmouth Steam Railway, Devon, 4th March 1989.

D3018   Derby   1953   81D   12/91   P   13018
08011   to Chinnor and Princes Risborough Railway, Oxfordshire, 25th April 1992.

D3019   Derby   1953   8J   7/73   P   D3019
despatched from Allerton Depot, November 1973; to Bescot Yard, where stored for three weeks; to Powell Duffryn Fuels Ltd, NCBOE Gwaun-cae-Gurwen Disposal Point, West Glamorgan, 13th December 1973; to BR Canton Depot, Cardiff, for repairs, 4th June 1978; to Gwaun-cae-Gurwen, 8th December 1978; to South Yorkshire Railway Preservation Society (HNRC), Meadowhall, Sheffield, 5th July 1990; to Battlefield Line, Shackerstone, 30th July 2001; to Tyseley, for wheel turning, November 2003; to Cambrian Railway Trust, Llynclys, Oswestry, 20th December 2003.

D3022   Derby   1953   41A   9/80   P   D3022
08015   to Severn Valley Railway (despatched from BR Swindon Works), 27th May 1983; to EWS Toton Depot, for repairs, 13th April 2006; returned to SVR, 14th April 2006.

D3023   Derby   1953   9D   5/80   P   08016
08016   to Hargreaves Industrial Services Ltd, NCBOE British Oak Disposal Point, Crigglestone, West Yorkshire, October 1980; to South Yorkshire Railway Preservation Society, Meadowhall, Sheffield, 24th January 1992; to Peak Rail, Rowsley, 4th April 2002; to Bluebell Railway, on hire, 27th March 2006; to Peak Rail, Rowsley, April 2008.

D3029   Derby   1953   15A   1/87   P   13029
08021   to Birmingham Railway Museum, Tyseley (despatched from BR Derby), 7th May 1987; to BR Tyseley Depot, for repairs, 3rd December 1988; returned to BRM.

D3030   Derby   1953   41A   3/85   P   LION
08022   to Guinness Ltd, Park Royal, London (despatched from BR Swindon Works, 5th July 1985), arrived 20th July 1985; to BR Old Oak Common, for repairs, April 1990; returned to Guinness; to BR Old Oak Common Depot, for repairs, August 1991; returned to Guinness; to RFS (Engineering) Ltd, Doncaster, for repairs, 13th December 1993; returned to Guinness, 11th January 1994; to Cholsey & Wallingford Railway, Oxfordshire, 31st August 1997.

D3038   Derby   1953   9A   12/72   F   2100/525
to NCB Ashington Central Workshops (despatched from BR Newton Heath), October 1973; to Bates Colliery, Blyth, March 1974; scrapped, June 1980.

D3044   Derby   1954   16A   8/74   P   33/ MENDIP
08032   to BR Derby Works, for overhaul, August 1974; to Foster Yeoman Quarries Ltd, Merehead, Somerset, February 1975; to BR Gloucester Depot, for repairs, 10th February 1980; returned to Merehead; to East Somerset Railway, Cranmore, for repairs, 10th August 2001; returned to Merehead, by 18th November 2001; to East Somerset Railway, for gala, April 2003; returned to Merehead; to East Somerset Railway, on loan, April 2003; returned to Merehead by October 2003; to East Somerset Railway, on loan, December 2004; returned to Merehead; to Whatley Quarry, about March 2006; returned to Merehead, 9th July 2006; to Mid-Hants Railway, Ropley, on loan, 15th August 2008; to Knight's Rail Services, Eastleigh Works, 19th October 2010.

D3059   Derby   1954   16F   5/80   P   13059/
08046   **BRECHIN CITY**
to Associated British Maltsters Ltd, Airdrie (despatched from BR Derby), 21st January 1981; to BR Motherwell Depot, for repairs, 7th February 1981; returned to Airdrie; to Caledonian Railway Ltd, Brechin, Angus, December 1985.

D3067   Darlington   1953   52A   2/80   P   201277/ M414
08054   to Tilcon Ltd, Swinden Lime Works, Grassington, about August 1980; to Embsay & Bolton Abbey Railway, 27th February 2008.

D3074   Darlington   1953   40A   6/84   P   UNICORN
08060   to Guinness Ltd, Park Royal, London (despatched from BR Swindon Works, 5th July 1985), arrived 20th July 1985; to BR Old Oak Common Depot, for repairs, early April 1988; returned to Guinness; to BR Old Oak Common Depot, for repairs, May 1989; returned to Guinness; to Cholsey & Wallingford Railway, Oxfordshire, 31st August 1997.

D3079   Darlington   1953   55B   12/84   P   13079
08064   to National Railway Museum, York, 28th October 1985; to National Railway Museum, Shildon, 18th August 2004; to National Railway Museum, York, 18th October 2005.

D3087   Derby   1954   8F   7/73   F   NFT
to CEGB Walsall Power Station, October 1973; to BR Tyseley Depot, November 1981; returned to CEGB Walsall; scrapped on site by Thos. W. Ward Ltd, May 1983.

D3088   Derby   1954   2F   12/73   F   D3088/ 2100-526
to NCB Ashington Colliery, 18th April 1974; to Ashington Workshops, 20th June 1974; to Bates Colliery, Blyth, 6th December 1974; to Lambton Engine Works, Philadelphia, 20th April 1979; to Bates Colliery, 6th September 1979; to Lambton Engine Works, 29th

June 1981; to Bates Colliery, 15th February 1983; scrapped on site, week ending 26th October 1985.

D3099   Derby   1955   73F   10/72   F   NFT
to Shipbreakers (Queenborough) Ltd, Kent, July 1973; used for spares; dismantled remains scrapped about 1980.

D3101   Derby   1955   73F   5/72   P   D3101
to ARC (East Midlands) Ltd, Loughborough, February 1973; to Great Central Railway, Loughborough, 14th December 1984.

D3102   Derby   1955   31A   11/77   P   007/ JAMES
08077   to Wiggins Teape & Co Ltd, Fort William, December 1978; to BR Eastfield Depot, Glasgow, for repairs, May 1979; returned to Wiggins Teape; to BR Glasgow Works, for repairs, June 1981; returned to Wiggins Teape; to BR Eastfield Depot, for repairs, 16th March 1984; returned to Wiggins Teape, 24th March 1984; sold to RFS (Engineering) Ltd, Kilnhurst, 23rd November 1990; to BREL, York, on hire, 16th December 1991; returned to RFS, 17th November 1992; to Roche Products, Dalry, on hire, 17th November 1992; to RFS Doncaster, about March 1993; to Teesbulk Handling, Middlesbrough, on hire, 2nd June 1994; returned to RFS, 13th September 1994; to Eastleigh, for storage, 10th December 1996; sold to Freightliner, and initially based at Southampton Docks.

D3110   Derby   1955   52A   3/86   F   08085
08085   to RFS (Engineering) Ltd, Doncaster, July 1988; to RFS Kilnhurst, July 1989; returned to RFS Doncaster, by 4th March 1990; used for spares; to C.F. Booth Ltd, Rotherham, for scrap, 24th March 1993; scrapped, 16th April 1993.

SECTIONS 15/16

Two variants of the popular shunter: Firstly British Railways built 0-6-0 diesel electric locomotives, numbered D3127-D3136, D3167-D3438, D3454-D3472, D3503-D3611, D3652-D3664, D3672-D3718, D3722-D4048, D4095-D4098, and D4115-D4192, introduced 1953. (Locomotives D3117-D3126 and D3152-D3166 were unclassified.) Fitted with an English Electric 6KT engine developing 350bhp at 680rpm, and driving wheels of 4ft 6in diameter. Later classified TOPS Class 08. Secondly British Railways built 0-6-0 diesel electric locomotives, numbered D3137-D3151, D3439-D3453, D3473-D3502, D3612-D3651 and D4049-D4094, introduced 1955. Fitted with a Lister-Blackstone ER6T engine developing 350bhp at 750rpm, and driving wheels of 4ft 6in diameter. Later classified TOPS Class 10.

D3167   Derby   1955   36A   4/88   P   D3167
08102   purchased by Lincoln City Council, 23rd March 1988; displayed at Central Station, Lincoln, from 23rd August 1988; to BR Doncaster Works, for overhaul, 12th October 1988; returned to Lincoln, 5th April 1989; displayed adjacent to Lincoln Station; to Great Northern & East Lincolnshire Railway, Ludborough, 8th May 1994.

D3174   Derby   1955   31A   7/84   P   D3174/
08108   DOVER CASTLE
to Dower Wood & Co Ltd, Newmarket, Suffolk, 1st August 1984; to East Kent Light Railway, Shepherdswell, Kent, 3rd August 1991; to Kent & East Sussex Railway, Tenterden, 13th October 1992.

D3179　Derby　　　　　　　1955　86A　　3/84　F　08113/ HO17
08113　to Powell Duffryn Fuels Ltd, NCBOE, Gwaun-cae-Gurwen (despatched from Canton Depot, Cardiff, 2nd August 1984), arrived 6th August 1984; sold to RMS Locotec Ltd, Dewsbury, 21st September 1995; to RMS Locotec, Wakefield, April 2006; to Morley Waste Traders Ltd, Leeds, February 2007; scrapped, May 2007.

D3180　Derby　　　　　　　1955　36A　　11/83　F　13180
08114　to Gloucestershire Warwickshire Railway, Toddington (despatched from BR Swindon Works, 2nd October 1984), arrived 3rd October 1984; to Swindon & Cricklade Railway, 14th November 1987; to Great Central Railway, Loughborough, 22nd January 1991; to Great Central Railway, Ruddington, Nottingham, 25th February 1997.

D3183　Derby　　　　　　　1955　82C　　12/72　F　D3183
to NCB Merthyr Vale Colliery, Aberfan, March 1973; to BR Canton Depot, Cardiff, for repairs, May 1975; to Merthyr Vale Colliery, June 1975; to BR Canton Depot, Cardiff, for repairs, 11th September 1980; to Merthyr Vale Colliery, 19th December 1980; to BR Canton Depot, Cardiff, for repairs, 19th July 1985; to Merthyr Vale Colliery, September 1986; scrapped by W. Phillips Ltd of Llanelli, December 1987.

D3190　Derby　　　　　　　1955　5A　　3/84　P　D3190/
08123　　　　　　　　　　　　　　　　　　　　　　**GEORGE MASON**
to Cholsey & Wallingford Railway, Oxfordshire (despatched from BR Swindon Works), 7th June 1985.

D3201　Derby　　　　　　　1955　40A　　9/80　P　08133
08133　to Sheerness Steel Co Ltd, Sheerness, Kent (despatched from BR Swindon Works), 6th October 1981; to RFS (Engineering) Ltd, Kilnhurst, for repairs, 29th March 1991; returned to Sheerness, 1st May 1991; to South Yorkshire Railway Preservation Society, Meadowhall, Sheffield, 30th August 1995; Barrow Hill Engine Shed Society, 19th December 2000; to Severn Valley Railway, 17th April 2002.

D3225　Darlington　　　　　1955　73F　　4/77　F　009
08157　to Independent Sea Terminals, Ridham Dock, Kent (despatched from BR Eastleigh Works), 28th July 1977; to RFS (Engineering) Ltd, Doncaster, 7th April 1993; to Coopers Metals Ltd, Sheffield, 28th May 1996; scrapped, 20th June 1996.

D3232　Darlington　　　　　1956　52A　　3/86　P　D3232
08164　to RFS (Engineering) Ltd, Doncaster (despatched from BR Tyne Yard), July 1988; while at RFS was named PRUDENCE; to RFS Kilnhurst, for weighing, 15th January 1991; returned to RFS, Doncaster, same day; to East Lancashire Railway, Bury, 28th January 1998.

D3236　Darlington　　　　　1956　55B　　3/88　P　13236
08168　to BREL Ltd, York, April 1989; loco included in sale when works privatised; sold at public auction, 25th June 1996; to Battlefield Line, Shackerstone, 12th July 1996; to Fragonset, Derby, 1st July 1999; to Alstom, Asfordby, 10th December 2001; to Battlefield Line, Shackerstone, 22nd July 2004; to Bluebell Railway, on hire, 30th April 2008.

D3238　Darlington　　　　　1956　52A　　3/86　F　08170
08170　to RFS (Engineering) Ltd, Doncaster (despatched from BR Tyne Yard), July 1988; to RFS Kilnhurst, 15th July 1989; used for spares, to C.F. Booth Ltd, Rotherham, 11th December 1991; scrapped, December 1991.

D3245 Derby 1956 55B 10/88 F D3245
08177 to BREL, Crewe, April 1989; loco included in sale when works privatised; scrapped on site by M.R.J. Phillips Ltd, August 1996.

D3255 Derby 1956 85B 12/72 P D3255
to NCB Blaenavon Colliery, 19th March 1973; to Bargoed Colliery, March 1973; to BR Canton Depot, Cardiff, for wheel turning, September 1974; returned to Bargoed Colliery; to BR Canton Depot, Cardiff, for repairs, 28th December 1975; to Bargoed Colliery, about April 1976; to BR Canton Depot, Cardiff, August 1976; to Bargoed Colliery, 1976; to BR Canton Depot, Cardiff, for repairs, 18th April 1977; to Bargoed Colliery, August 1977; to BR Canton Depot, Cardiff, March 1978; to Bargoed Colliery, about May 1978; to BR Canton Depot, Cardiff, 23rd May 1979; to Bargoed Colliery, about August 1979; to BR Canton Depot, Cardiff, 6th April 1981; to Mardy Colliery, 27th May 1981; to Mountain Ash Colliery, 28th December 1981; to Mardy Colliery, 27th May 1982; to Brighton Railway Museum, about May 1987; to Colne Valley Railway, Castle Hedingham, 9th September 2008; to Tim Ackerley, 27th August 2009 (stored at a private farm at East end of York).

D3261 Derby 1956 86A 12/72 P D3261
to NCB Tower Colliery, Hirwaun, Glamorgan, July 1973; to BR Canton Depot, Cardiff, for repairs, September 1975; to Tower Colliery, October 1975; to BR Canton Depot, Cardiff, 27th October 1977; to Tower Colliery, 2nd November 1977; to BR Canton Depot, Cardiff, 28th October 1978; to BR Swindon Works, April 1979; to Tower Colliery, 29th November 1979; to Brighton Railway Museum, 11th December 1988; to Swindon & Cricklade Railway, about March 1996.

D3265 Derby 1956 86A 9/83 P 13265/ MARK
08195 to Llangollen Railway (despatched from BR Swindon Works), 25th March 1986.

D3272 Derby 1956 86A 5/89 P 08202/CHUFFER
08202 to G.G. Papworth Ltd, Ely, Cambridgeshire, 10th March 1990; to The Potter Group Ltd, Knowsley, Merseyside, 9th January 2001; to Gloucestershire Warwickshire Railway, for gala and then storage, 7th July 2010; to GBRf, Celsa Steelworks, Cardiff, on hire, late January 2011; to The Potter Group, Ely, 4th May 2011.

D3286 Derby 1956 16C 11/80 F 08216
08216 to Sheerness Steel Co Ltd, Sheerness (despatched from BR Swindon Works, 22nd April 1983), arrived 12th May 1983; to RFS (Engineering) Ltd, Kilnhurst, for overhaul, 21st December 1989; returned to Sheerness, 5th May 1990; dismantled by June 1995; to South Yorkshire Railway Preservation Society (HNRC), Meadowhall, Sheffield, 2nd April 1996; to Barrow Hill Engine Shed Society, 10th April 2001; scrapped, 19th April 2001.

D3290 Derby 1956 5A 3/86 P 08220
08220 to William Smith Ltd, Wakefield, West Yorkshire (despatched from BR Chester), 11th July 1988; to Steamtown, Carnforth, Lancashire, 12th July 1990; to Wrenbury Station, Cheshire, April 2006; to Great Central Railway, Ruddington, October 2008.

D3308 Darlington 1956 85B 3/84 P 13308/ CHARLIE
08238 sold to Forest Free Mining, Tetbury, 1984; intended for Parkend Mine, Forest

of Dean, but stored at BR Gloucester Depot until November 1988; to BR Swindon Works, 11th November 1988; to Swindon Heritage Centre, 29th November 1988; to Dean Forest Railway, Lydney, January 1993; to RFS (Engineering) Ltd, Doncaster, for repairs, December 1997; returned to Dean Forest Railway, 1998.

D3309   Darlington   1956   55H   7/84   F   3309
08239   to C.F. Booth Ltd, Rotherham, 1995; resold to South Yorkshire Railway Preservation Society (HNRC), Meadowhall, Sheffield, 12th November 1998; to European Metal Recycling Ltd, Kingsbury, about June 2001; scrapped, October 2005.

D3336   Darlington   1957   41J   3/85   P   08266
08266   to Keighley & Worth Valley Railway, Haworth, West Yorkshire (despatched from BR Swindon Works), 21st November 1985.

D3342   Derby   1957   31B   7/87   F   08272
08272   to RFS (Engineering) Ltd, Doncaster, South Yorkshire, April 1988; used for spares only; scrapped, January 1991.

D3358   Derby   1957   82C   1/83   P   D3358
08288   to Mid-Hants Railway, Ropley, Hampshire (despatched from BR Swindon Works, 13th September 1984), arrived 1st November 1984; to Wabtec, Doncaster, for repairs, 20th August 2002; returned to Mid-Hants Railway, Ropley, October 2002.

D3362   Derby   1957   65A   5/84   F   08292
08292   to Deanside Transit Ltd, Glasgow, July 1984; scrapped on site, November 1994.

D3366   Derby   1957   55B   10/88   P   08296/ 001
08296   to BREL, Crewe, April 1989; loco included in sale when works privatised; to ABB British Wheelset Ltd, Trafford Park, Manchester, January 1995; to ABB Transportation, Crewe, August 1996; to East Somerset Railway, Cranmore, for storage, February 2000; to Foster Yeoman Quarries Ltd, Merehead, about March 2000; to Crewe Works, for open day, May 2000; returned to Merehead; to Hanson Aggregates, Whatley Quarry, by 16th September 2001; to Foster Yeoman, Isle of Grain, 28th June 2002; to Foster Yeoman, Whatley Quarry, 29th July 2002; to Merehead, June 2003; to Isle of Grain, 2004; to Acton Rail Terminal, late 2004; to Isle of Grain, February 2005; to Merehead, by 13th July 2005; to Hanson Aggregates, Machen Quarry, on hire, 6th April 2006. (It has been suggested in the railway press that this locomotive may in fact be the body of 08787 mounted on the frames of 08296 – and that a locomotive with the body of 08296 was scrapped at C.F. Booth Ltd, Rotherham, May 1994.)

D3378   Derby   1957   36A   11/92   P   08308/ LANGWITH
08308   to South Yorkshire Railway Preservation Society, Meadowhall, Sheffield, 15th July 1993; purchased by R.T. Rail, Crewe, 1st May 1998; to RFS, Doncaster, for repairs, 21st January 1999; to Scot Rail, Inverness, on hire, 7th September 1999; to Wabtec, Doncaster, for repairs, 16th November 2005; to Scot Rail, Inverness, on hire, 4th May 2006; RMS Locotec acquired RT Rail, November 2007.

D3390   Derby   1957   16A   12/82   F   P400D/ SUSAN
08320   sold to Forest Free Mining, Tetbury (despatched from Toton), 2nd October 1984; intended for Parkend Mine, Forest of Dean, but stored at Gloucester Depot from 5th October 1984 until December 1988; to English China Clays Ltd, Blackpool Driers,

Burngullow, Cornwall, 2nd December 1988; to Imerys, Rocks Driers, Bugle, about January 2009; to European Metal Recycling, Kingsbury, for scrap, 28th July 2010; scrapped October 2010.

D3401　　Derby　　　　　　　　1957　　36A　　4/88　　P　08331/ H001/
08331　　　　　　　　　　　　　　　　　　　　　　　　　　　　TERENCE
to RFS (Engineering) Ltd, Doncaster, April 1988; to RFS Kilnhurst, 5th August 1991; to Inco Europe, Clydach, Swansea, on hire, 17th September 1991; returned to RFS Kilnhurst, 6th November 1991; to Channel Tunnel contract, on hire, 23rd September 1992; to RFS Kilnhurst, 5th February 1993; to Flixborough Wharf, Scunthorpe, on hire, 8th March 1993; to Allied Steel & Wire Ltd, Cardiff, on hire, 24th September 1994; returned to RFS Doncaster, 27th November 1994; to Great North Eastern Railway, Craigentinny, Edinburgh, on hire, 30th January 1997; returned to RFS, Doncaster, by February 1998; to Hays Chemicals, Sandbach, on hire, August 1998; returned to RFS, September 1998; to GNER, Craigentinny, Edinburgh, on hire, 12th October 1998; to Wabtec, Doncaster, for repairs, December 2000; returned to Craigentinny; to Wabtec, Doncaster, 11th August 2005; purchased by R.T. Rail, 7th May 2006; to Embsay & Bolton Abbey Railway, for testing, May 2007; acquired by RMS Locotec, November 2007; to Wabtec, Doncaster, by 5th December 2007; to Lafarge Aggregates, Barrow upon Soar, on hire, 13th December 2007; to Midland Railway, Butterley, for storage, by 5th July 2008; to Cemex Rail Products, Washwood Heath, on hire, by 17th November 2010.

D3405　　Derby　　　　　　　　1957　　41A　　1/87　　F　08335
08335　　to Thomas Hill (Rotherham) Ltd, Kilnhurst, South Yorkshire, 20th November 1987; to RFS (Engineering) Ltd, Kilnhurst, July 1989; scrapped on site by C.F. Booth Ltd, Rotherham, 26th July 1989.

D3407　　Derby　　　　　　　　1957　　36A　　2/87　　F　08337
08337　　to RFS (Engineering) Ltd, Doncaster, April 1988; used for spares; remains scrapped, January 1989.

D3415　　Derby　　　　　　　　1958　　67C　　11/83　F　RUSSELL
08345　　to Deanside Transit Ltd, Glasgow, May 1985; to HNRC, Long Marston, 21st June 2007; to C.F. Booth Ltd, Rotherham, for scrap, 4th November 2009; scrapped 13th November 2009.

D3420　　Crewe　　　　　　　　1957　　86A　　1/84　　F　D3420
08350　　to Churnet Valley Railway, Cheddleton (despatched from BR Swindon Works, 17th September 1984), arrived 18th September 1984; to LNWR Ltd, Carriage Works, Crewe, July 2004; to Midland Railway, Butterley, 5th September 2007; to Heanor Haulage, Langley Mill, for storage, early September 2008; to Ron Hull Ltd, Rotherham, for scrap, week ending 19th September 2008; scrapped, March 2010.

D3429　　Crewe　　　　　　　　1958　　86A　　1/84　　P　D3429
08359　　to Churnet Valley Railway, Cheddleton (despatched from BR Swindon Works, 19th September 1984), arrived 20th September 1984; to Peak Rail, Buxton, 9th January 1987; to Peak Rail, Darley Dale, Derbyshire, December 1989; to Battlefield Line, Shackerstone, 16th October 1996; to Birmingham Railway Museum, Tyseley, 29th June 1999; to Northampton & Lamport Railway, Chapel Brampton, 11th August 2005; to Telford Steam Railway, 20th January 2007; to Chasewater Railway, on loan, 19th July 2010.

D3452 Darlington 1957 16A 6/68 P D3452
to ECC Ports Ltd, Fowey, Cornwall, September 1968; to Bodmin & Wenford Railway, Cornwall, 5th March 1989.

D3460 Darlington 1957 86A 11/91 P 08375/21
08375 to Railway Age, Crewe, 13th August 1993; to LNWR Ltd, Carriage Works, Crewe, about January 1998; purchased by R.T. Rail, Crewe; to RMS Locotec Ltd, Dewsbury, for overhaul, 25th March 1998; to Direct Rail Services, Sellafield, on hire, about August 1998; to RFS (Engineering) Ltd, Doncaster, for repairs, about January 1999; to Port of Felixstowe, on hire, 2nd November 1999; to Freightliner, Ipswich, on hire, by 21st November 1999; to Wabtec, Doncaster, for repairs, December 2001; to Port of Felixstowe, on hire, by 24th July 2001; to Flixborough Wharf, on hire, January 2002; to Wabtec, Doncaster, May 2002; to Hays Chemicals, Sandbach, on hire, 5th July 2002; returned to Wabtec, for repairs, 2nd December 2002; to Hays Chemicals, Sandbach, on hire, about January 2003; to Wabtec, Doncaster, March 2003; to Port of Felixstowe, on hire, 5th June 2003; to Alstom, Eastleigh Works, on hire, 18th November 2003; to Hays Chemicals, Sandbach, on hire, September 2004; to Wabtec, Doncaster, for repairs, 2005; returned to Hays, on hire, 15th September 2005; to PD Ports, Tees Dock, on hire, 15th October 2005; to Wabtec, Doncaster, for repairs, 15th February 2006; to Manchester Ship Canal, Trafford Park, on hire, about May 2006; to Wabtec, December 2006; to DHL, ProLogis Park Industrial Estate, Coventry, on hire, 5th February 2007; to Corus, Trostre Works, on hire, 18th December 2009; to Castle Cement, Ketton, on hire, November 2010; to PD Ports, Tees Dock, on hire, 28th February 2011.

D3462 Darlington 1957 84A 6/83 P D3462
08377 to Dean Forest Railway (despatched from BR Swindon Works), 19th March 1986; to Rail & Marine Engineering, Thingley Junction, Chippenham, for storage, 1995; to West Somerset Railway, Minehead, 23rd April 1996.

D3476 Darlington 1957 16A 6/68 F D3476
to ECC Ports Ltd, Fowey, Cornwall, 6th September 1968; to storage in the Midlands, 5th March 1989; to South Yorkshire Railway Preservation Society (HNRC), Meadowhall, Sheffield, 2nd October 1989; to Colne Valley Railway, Essex, 1st December 2000; to hauliers yard (possibly Wishaw) where stripped for spares, 26th February 2009; to T.J. Thomson Ltd, Stockton, 3rd March 2009; scrapped March 2009.

D3489 Darlington 1958 16A 4/68 P D3489
to Felixstowe Dock & Railway Co Ltd, Suffolk, August 1968; to BR Swindon Works, for repairs, 30th January 1980; to Port of Felixstowe, 19th May 1980; to BR Stratford Depot, for repairs, 11th October 1984; to Port of Felixstowe, December 1984; to Wilmott Bros, Ilkeston, for repairs, 27th July 1990; to Felixstowe, 18th September 1990; to Spa Valley Railway, Tunbridge Wells, 17th August 2001.

D3497 Doncaster 1957 16B 4/68 F D3497
to ECC Ports Ltd, Fowey, Cornwall, 21st August 1968; dismantled (as a source of spares for D3452 and D3476), 1988; remains scrapped, February 1990.

D3503 Derby 1958 40B 6/96 F 08388
08388 sold to Mike Darnall, Newton Heath, Manchester, about 1999; to Wabtec, Doncaster, for repairs, 1999; returned to Darnall about July 2000; to European Metal Recycling, Kingsbury, for scrap, 15th July 2010; scrapped, early October 2010.

D3505   Derby   1958   86A   3/93   F   08390
08390   to South Yorkshire Railway Preservation Society (HNRC), Meadowhall, Sheffield (despatched from Adtranz, Crewe), 4th April 1997; used for spares; to Barrow Hill Engine Shed Society, Staveley, 6th March 2001; scrapped at Barrow Hill by HNRC, March 2004.

D3513   Derby   1958   82A   7/85   F   402D/ ANNABEL
08398   to ECC Ports Ltd, Fowey, Cornwall, 3rd February 1986; to BR Laira Depot, Plymouth, for repairs, summer 1988; to ECC Marsh Mills, Devon, July 1988; to BR Laira Depot, for repairs, 29th September 1992; returned; to English China Clays Ltd, Rocks Driers, near Bugle, about March 1993; to European Metal Recycling, Kingsbury, for scrap, 9th September 2010; scrapped, January 2011.

D3516   Derby   1958   40B   2/04   P   08401
08401   to LH Group, Barton under Needwood, 27th January 2011; repaired; to GBRf, Whitemoor Yard, March, on hire, 18th March 2011; to GBRf, Cardiff, on hire, 6th April 2011; to GBRf, Whitemoor Yard, March, on hire, 26th May 2011.

D3526   Derby   1958   66B   4/04   P   08411
08411   to Traditional Traction, Colne Valley Railway (despatched from EWS Allerton Depot), 29th March 2007.

D3528   Derby   1958   41A   12/96 F   08413
08413   to C.F. Booth Ltd, Rotherham (despatched from Wabtec, Doncaster), 13th June 2000; sold to RMS Locotec, Dewsbury, 3rd June 2004; to Morley Waste Traders Ltd, Leeds, for scrap, February 2007; scrapped, February 2007.

D3529   Derby   1958   81A   7/99   F   08414
08414   to Traditional Traction, Wishaw, Warwickshire (despatched from EWS Toton Depot), 9th March 2007; to European Metal Recycling, Kingsbury, for scrap, 16th August 2007; scrapped, late August 2007.

D3530   Derby   1958   8F   10/96 F   08415
08415   to RFS (Engineering) Ltd, Doncaster, 30th September 1996; used for spares; to Coopers (Metals) Ltd, Sheffield, for scrap, 5th September 1997; scrapped, 1997.

D3531   Derby   1958   16A   2/92   F   08416
08416   to RFS, Kilnhurst, 10th April 1992; used for spares; remains scrapped on site at Kilnhurst, August 1993.

D3532   Derby   1958   NTW   —-   P   08417
08417   owned by Serco Railtest and allocated to Etches Park Depot, Derby; to Foster Yeoman Quarries Ltd, Merehead, on hire, 17th June 1999; returned to Serco Railtest, 30th March 2000; to Hanson, Somerset, on hire, 2000; to Foster Yeoman, Merehead, on hire, by 16th September 2001; to Serco Railtest, Derby, 9th April 2002.

D3533   Derby   1958   16A   2/04   P   08418
08418   to West Coast Railway Company, Carnforth (despatched from Bescot), 4th August 2010.

D3534   Derby   1958   12B   4/93   F   08419
08419   to Mike Darnall, Newton Heath, Manchester (despatched from Crewe Works), and moved to Bombardier Transportation, Doncaster Works, for storage, 23rd November 2000; to C.F. Booth Ltd, Rotherham, for scrap, 24th November 2004; scrapped, November 2004.

D3538   Derby   1958   8F   11/88   P   08423/ 14 / HO11
08423   to Trafford Park Estates Ltd, Manchester, 21st November 1988; to MoD Kineton, for trials, 1st August 1994; sold to RMS Locotec Ltd, Dewsbury, 8th August 1994; to Mobil Oil Co Ltd, Coryton, on hire, about July 1995; to Faber Prest Ports Ltd, Flixborough Wharf, on hire, about June 1998; returned to RMS, 3rd July 2003; to Faber Prest Ports Ltd, Flixborough Wharf, on hire, 18th February 2004; suffered fire damage, 11th September 2006; to RMS Locotec, Wakefield, for repairs, 10th October 2006; to PD Ports, Tees Dock, Grangetown, on hire, week ending 23rd February 2007; to Flixborough Wharf, on hire, late 2007; returned to PD Ports, on hire, by April 2008.

D3551   Derby   1958   36A   1/92   P   08436
08436   to South Yorkshire Railway Preservation Society, Meadowhall, Sheffield, late May 1993; purchased by R.T. Rail, Crewe, about April 1998; to RMS Locotec Ltd, Dewsbury, for overhaul, 1st May 1998; to Keighley & Worth Valley Railway, Haworth, for painting, 1998; to Hays Chemicals, Sandbach, on hire, 8th January 1999; to Railway Age, Crewe, for repairs, 17th December 2000; returned to Hays Chemicals, 22nd December 2000; to Wabtec, Doncaster, for repair, 26th September 2001; returned to Hays Chemicals, on hire, by May 2002; returned to R.T. Rail, November 2002; to Hays Chemicals, Sandbach, on hire, March 2003; to Wabtec, Doncaster, by July 2003; to Swanage Railway, August 2004.

D3556   Derby   1958   66B   2/04   P   08441
08441   to Traditional Traction, Wishaw, Warwickshire (despatched from EWS Motherwell Depot), 14th May 2007; to Colne Valley Railway, June 2007; to LH Group, Barton under Needwood, for repairs, 16th February 2010; to Colne Valley Railway, November 2010; to Port of Felixstowe, on hire, 7th February 2011.

D3558   Derby   1958   62A   7/85   P   08443
08443   to Scottish Grain Distillers, Cambus Distillery, Alloa (despatched from BR Grangemouth), 14th January 1986; to Scottish Railway Preservation Society, Bo'ness, 28th June 1993.

D3559   Derby   1958   86A   11/86   P   08444
08444   to Bodmin & Wenford Railway, Cornwall, 27th March 1987.

D3560   Derby   1958   40B   11/95   P   08445
08445   purchased by Mike Darnall, Newton Heath, Manchester, about 1999; to Wabtec, Doncaster, for repairs, 1999; returned to Darnall, 17th July 2000; to East Lancashire Railway, Bury, by 14th September 2001; to Carillion Construction Ltd, Manchester Metrolink upgrade, on hire, 26th May 2007; to former Corus Works, Castleton, for storage, from about 31st August 2007; to Castle Cement, Ketton, on hire, 18th February 2009; to Corus, Shotton Steelworks, on hire, 22nd July 2009; to LH Group, Barton under Needwood, for repairs, 27th March 2011; to GBRf, Barton Dock, Trafford Park, on hire, April 2011; to Daventry International Rail Freight Terminal, on hire, week ending 24th June 2011; to LH Group, Barton under Needwood, by 14th August 2011.

D3562   Derby   1958   12B   6/95   P   08447
08447   to Deanside Transit Ltd, Glasgow, about May 1995; to HNRC, Long Marston, 12th June 2007; overhauled; to Deanside Transit Ltd, Glasgow, 13th June 2008.

D3575   Crewe   1958   8J   4/04   P   08460
08460   to Traditional Traction, Colne Valley Railway (despatched from EWS Allerton Depot), 29th March 2007; to Port of Felixstowe, on hire, 5th November 2009; to Colne Valley Railway, for repairs,13th January 2011.

D3585   Crewe   1958   5A   3/86   F   08470
08470   to BREL Ltd, Crewe, April 1989; loco included in sale when works privatised; scrapped on site by M.R.J. Phillips Ltd, August 1996.

D3586   Crewe   1958   15A   9/85   P   D3586
08471   to Severn Valley Railway (despatched from BR Swindon Works), 14th April 1986.

D3587   Crewe   1958   NTW   —-   P   08472
08472   to Great North Eastern Railway and allocated to Bounds Green Depot; sold to RFS (Engineering) Ltd, Doncaster, 1998, the loco remaining at Bounds Green, on hire; to GNER, Craigentinny, on hire, 1999; returned to Wabtec, Doncaster, December 2000; to GNER, Craigentinny, on hire, by August 2003; returned to Wabtec, by 1st April 2004; to GNER, Craigentinny, on hire, 10th August 2005; to Wabtec, 26th November 2010.

D3588   Crewe   1958   86A   3/86   P   08473
08473   to T.J. Thomson Ltd, Stockton (despatched from Leicester), 28th November 2000; partly scrapped; remains sold to Dean Forest Railway, Lydney, for spares, 5th March 2001.

D3591   Crewe   1958   62A   9/85   P   D3591
08476   to Swanage Railway, Dorset (despatched from BR Swindon Works), 21st March 1986; to Battlefield Line, Shackerstone, 24th July 2009.

D3594   Horwich   1958   86A   11/91   P   08479
08479   to East Lancashire Railway, Bury, 30th April 1993.

D3596   Horwich   1958   86A   4/02   F   08481
08481   to Barry Island Railway, Barry Island (despatched from Wigan), 7th October 2005; to European Metal Recycling, Kingsbury, for scrap, May 2011; scrapped early June 2011.

D3599   Horwich   1958   ZN   6/95   P   08484
08484   to Railcare Ltd, Wolverton Works, June 1995; loco included in sale when works privatised; to LH Group, Barton under Needwood, for repair, 16th November 2003; returned to Wolverton Works, 10th February 2004; sold to Traditional Traction, April 2006; to Port of Felixstowe, on hire, 25th April 2006; to Traditional Traction, Wishaw, Warwickshire, for storage, about 10th November 2006; to Nene Valley Railway, for gala, February 2007; returned to Traditional Traction, February 2007; to St Phillips Marsh Depot, Bristol, for tyre turning, 19th April 2007; to Port of Felixstowe, on hire, week ending 4th May 2007.

D3600   Horwich   1958   12A   11/07 P 08485
08485   to West Coast Railway Company, Carnforth, 1st September 2010.

D3605   Horwich   1958   62A   11/85 P 08490
08490   to Strathspey Railway, Aviemore (despatched from BR Perth), 18th June 1987.

D3607   Horwich   1958   66B   ?   P 08492
08492   to Barrow Hill Engine Shed Society (HNRC), Staveley (despatched from EWS Motherwell Depot), 2nd to 5th June 2006.

D3608   Doncaster   1958   86A   7/99   F 08493
08493   sold to R.T. Rail of Crewe (despatched from Canton Depot, Cardiff), and moved direct to Wabtec, Doncaster, for storage, September 2003; to C.F. Booth Ltd, Rotherham, for scrap, 19th May 2008; scrapped, August 2008.

D3613   Darlington   1958   40A   2/69   F DAVID
to NCB Bestwood Colliery, Nottinghamshire, August 1969; to Linby Colliery, by September 1971; to Moor Green Colliery, Newthorpe, November 1971; to BR Toton Depot, for repairs, by 26th October 1975; returned to Moor Green Colliery, by 22nd December 1975; scrapped on site by The Vic Berry Company of Leicester, 10th April 1985.

D3618   Darlington   1958   40A   4/69   F ROBIN/ D16
to NCB Bestwood Colliery, Nottinghamshire, August 1969; to Annesley Colliery, March 1970; to BR Toton Depot for repairs, June 1974; returned to Annesley Colliery; to Cotgrave Colliery, 24th July 1980; to Moor Green Colliery, Newthorpe, 30th March 1981; scrapped on site by The Vic Berry Company of Leicester, 26th March 1985.

D3619   Darlington   1958   40A   2/69   F SIMON/ D15
to NCB Gedling Colliery, Nottinghamshire, September 1969; to Bestwood Colliery, November 1969; to Linby Colliery, about July 1971; to Moor Green Colliery, Newthorpe, 24th November 1975; scrapped on site by The Vic Berry Company of Leicester, week ending 5th April 1985.

D3638   Darlington   1958   52A   11/70 F 9185-61
to NCB Bates Colliery, Blyth, on hire, 19th November 1970; sold to NCB, 28th January 1971; to BR Gateshead Depot, for repairs, 4th February 1971; to BR Cambois Depot, Blyth, for repairs, 12th February 1971; to Bates Colliery, 16th February 1971; to Ashington Colliery, 15th April 1971; to Bates Colliery, 22nd April 1971; to Ashington Central Workshops, March 1975; used for spares; remains scrapped, September 1975.

D3642   Darlington   1958   36C   6/69   F 37
to BSC Redbourn Works, Scunthorpe, September 1969; to BSC Appleby-Frodingham Works, Scunthorpe, October 1975; scrapped, October 1978.

D3648   Darlington   1959   52A   1/71   F 9185-60
to NCB Bates Colliery, Blyth, on hire, 16th February 1971; to BR Cambois Depot, Blyth, for repairs, 27th February 1971; to Bates Colliery, 1st March 1971; sold to NCB, 1st March 1971; to BR Cambois Depot, for further repairs, 5th March 1971; to Bates Colliery, 23rd March 1971; scrapped on site by L. Marley & Co Ltd of Stanley, February 1977.

**D3654	Doncaster	1958	81A	6/04	P	08499**
08499	to Pullman Rail, Canton Depot, Cardiff (loco included in sale when depot sold), June 2005.

**D3657	Doncaster	1958	51L	10/88	P	08502/**
**08502	LYBERT DICKINSON**
to ICI Ltd, Wilton Works, Middlesbrough, 6th September 1988; sold to HNRC, February 2007; to Barrow Hill Engine Shed Society, Staveley, for repaint, 3rd August 2007; to Northern Rail Ltd, Heaton Traincare Depot, on hire, 10th September 2007.

**D3658	Doncaster	1958	55A	10/88	P	08503**
08503	to ICI Ltd, Wilton Works, Middlesbrough (despatched from BR Doncaster Works), 26th September 1988; sold to HNRC, February 2007; to Heanor Heavy Haulage, Langley Mill, for storage, 7th December 2007; to Moveright International, Wishaw, for storage, late 2008.

**D3662	Doncaster	1958	81D	8/93	P	08507**
08507	to S.M. McGregor & Sons, Bicester, for store, about April 1994; to South Yorkshire Railway Preservation Society (HNRC), Meadowhall, Sheffield, 23rd June 1995; to Barrow Hill Engine Shed Society, Staveley, 3rd November 1999; to Bombardier Transportation, Central Rivers Depot, Barton under Needwood, on hire, 7th April 2001; to Barrow Hill Engine Shed Society, Staveley, for repairs, 10th January 2011; to Port of Boston, on hire, 13th June 2011.

**D3673	Darlington	1958	66B	3/04	P	08511**
08511	to Traditional Traction, Wishaw (despatched from EWS Ayr Depot), 29th June 2007; to Vale of Glamorgan Railway, Barry Island, 9th July 2007; to Port of Felixstowe, on hire, 8th July 2009; to Wembley, for tyre turning, 2nd July 2010; to Colne Valley Railway, for painting, 2nd August 2010; to Port of Felixstowe, on hire, 12th January 2011.

**D3677	Darlington	1958	52A	?	F	08515**
08515	to Gwent Demolition, 1995; firm went bankrupt and sale cancelled; resold to T.J. Thomson, Stockton-on-Tees, for scrap, April 2000; dismantled at Gateshead TMD, February 2001; rolling chassis only to T.J. Thomson, Stockton-on-Tees, February 2001; to Foster Yeoman, Merehead, for spares, 2001; remains scrapped, 2001.

**D3679	Darlington	1958	30A	9/93	F	08517**
08517	to Barrow Hill Engine Shed Society, Staveley (HNRC); despatched from BR Stratford, 4th June 2001; to Wabtec, Doncaster, 27th June 2001; rebuilt using spares from 08668; stored in West Yard, Doncaster, late 2002; to C.F. Booth Ltd, Rotherham, 2007, but not scrapped: to HNRC, Long Marston, 19th December 2007; stripped for spares, early 2011; to C.F. Booth Ltd, Rotherham, for scrap, 1st July 2011; scrapped, 21st July 2011.

**D3685	Doncaster	1958	?	?	P	08523**
08523	to R.T. Rail, Crewe, and stored at Crewe Diesel Depot, 2004; to LNWR Ltd, Carriage Works, Crewe, for overhaul, 4th March 2004; to Heritage Centre, Crewe, by July 2005; to Hays Chemicals, Sandbach, on hire, about July 2005; to RMS Locotec, Wakefield, for repairs, February 2007; to Celtic Energy, Onllwyn Disposal Point, on hire, 30th March 2007; resold to RMS Locotec, Wakefield, and moved to their yard, about November 2007; to Onllwyn Disposal Point, on hire, by February 2008.

D3689   Darlington   1959   ZI   6/95   P   08527
08527   to ABB (Customer Support) Ltd, Ilford, June 1995; loco included in sale when works privatised; to Barrow Hill Engine Shed Society, Staveley (HNRC), September 2006; to Fastline, Roberts Road Depot, Doncaster, on hire, 7th June 2007; to Flixborough Wharf, on hire, 31st August 2010.

D3690   Darlington   1959   2F   7/05   P   08528
08528   to Battlefield Line, Shackerstone, 31st July 2010.

D3691   Darlington   1959   40B   ?   F   08529
08529   to R.T. Rail, Crewe (despatched from Doncaster Depot), April 2005; to Wabtec, Doncaster, for repairs, July 2005; to RMS Locotec, Dewsbury, August 2005; to Wabtec, for storage, by November 2005; to C.F. Booth Ltd, Rotherham, for scrap, 21st May 2008; scrapped, August 2008.

D3692   Darlington   1959   ?   ?   P   08530
08530   to Traditional Traction, Wishaw, Warwickshire, about August 2006; to Port of Felixstowe, on hire, about January 2007; to LH Group, Barton under Needwood, for repairs, 24th December 2007; sold to Freightliner, March 2009, and based initially at Southampton Docks.

D3699   Darlington   1959   2F   10/00   F   08535
08535   to R.T. Rail, Crewe (despatched from Crewe Diesel Depot), and stored at LNWR Ltd, Carriage Works, Crewe, 4th March 2004; to RMS Locotec, Wakefield, for repairs, 5th September 2007; sold to RMS Locotec, November 2007; to Corus, Shotton Steelworks, 21st January 2009; dismantled for spares; to C.F. Booth Ltd, Rotherham, for scrap, 15th September 2009; scrapped, 17th September 2009.

D3700   Darlington   1959   16C   6/95   P   08536
08536   to Rail Vehicle Engineering Ltd, Derby (despatched from East Midlands Trains, Etches Park, Derby), July 2010.

D3723   Darlington   1959   1A   7/90   P   08556
08556   to North Yorkshire Moors Railway, 23rd October 1993.

D3735   Crewe   1959   ZH   6/95   P   08568/ ST ROLLOX
08568   to Railcare Ltd, Glasgow Works, June 1995; loco included in sale when works privatised.

D3738   Crewe   1959   NTW   —-   P   08571
08571   to Great North Eastern Railway and allocated to Craigentinny Depot, Edinburgh; sold to RFS (Engineering) Ltd, Doncaster, about March 1997; to GNER, Craigentinny, on hire, about April 1997; returned to RFS, 1998; to ARC, Whatley, on hire, February 1999; returned to RFS, Doncaster; to Hanson, Whatley Quarry, on hire, by 6th January 2001; to Merehead, September 2001; to Wabtec, Doncaster, June 2002; to GNER, Bounds Green, on hire, about March 2005.

D3740   Crewe   1959   ZI   6/95   P   08573
08573   to ABB (Customer Support) Ltd, Ilford, June 1995; loco included in sale when works privatised; sold to R.T. Rail, Crewe, about January 2001; to Wabtec, Doncaster, for repairs, 25th October 2001; to Channel Tunnel Rail Link, Beechbrook Farm, near Ashford, on hire, 9th November 2001; to Wabtec, Doncaster, for repairs, by 10th January

2003; to Bombardier Transportation, Ilford, on hire, by September 2003; to Wabtec, Doncaster, for repairs, August 2004; to Freightliner, Coatbridge, on hire, 12th March 2005; to Scot Rail, Inverness, on hire, April 2005; to Wabtec, 8th May 2006; to Tubelines, Ruislip, London, on hire, 17th May 2006; sold to RMS Locotec, Wakefield, November 2007; to Wabtec, for repairs, about May 2007; to Bombardier Transportation, Ilford, on hire, 12th July 2007.

D3743   Crewe   1959   86A   6/00   F   08576
08576   to Battlefield Line, Shackerstone (despatched from Canton Depot, Cardiff), 12th February 2004; used for spares; to T.J. Thomson Ltd, Stockton, for scrap, 8th June 2006; scrapped, May 2007.

D3755   Crewe   1959   55H   ?   P   08588/17
08588   to RMS Locotec, Dewsbury (despatched from Neville Hill Depot direct to Wabtec, Doncaster, for repairs), 16th April 2005; sold to Wabtec, Doncaster, May 2005; to Network Rail, Whitemoor Yard, March, on hire, 15th September 2005; sold to RT Rail, about March 2006; to Wabtec, for repairs, 31st March 2006; to RMC Aggregates, Dove Holes, on hire, 7th April 2006; to CTRL, Ripple Lane, on hire, 28th September 2006; to Wabtec, for repairs, 6th November 2006; to Bombardier Transportation, Ilford, on hire, 20th November 2006; sold to RMS Locotec, Wakefield, November 2007; to Network Rail, Whitemoor Yard, March, on hire, December 2007; to PD Ports, Tees Dock, on hire, July 2008; to Weardale Railway, Wolsingham, 1st March 2011.

D3757   Crewe   1959   52B   9/93   P   08590/RED LION
08590   to Midland Railway, Butterley, Derbyshire, July 1994.

D3761   Crewe   1959   16A   2/97   F   08594
08594   to Mike Darnall, Newton Heath, Manchester, November 2000; to Wabtec, Doncaster, by April 2001; to Mike Darnall, 9th January 2008; to Traditional Traction, Wishaw, 15th July 2010; to European Metal Recycling, Kingsbury, for scrap, 21st July 2010; scrapped, September 2010.

D3763   Derby   1959   81A   3/77   P   08596/ H0006
08596   to Bowaters UK Paper Co Ltd, Sittingbourne, Kent, 16th May 1977; to BR Swindon Works, for overhaul, 20th November 1981; returned to Bowaters, 4th January 1982; sold to RFS (Engineering) Ltd, Kilnhurst, 6th June 1991; to ECC, Quidhampton, on hire, 29th July 1991; returned to RFS, Doncaster, 5th to 7th September 1991; to Channel Tunnel, Folkestone, (Balfour Beatty Ltd), on hire, 21st September 1991; to RFS, Doncaster, 6th July 1993; to Sheerness Steel Co Ltd, Sheerness, on hire, July 1993; returned to RFS, Doncaster, by February 1997; to EWS, Decoy Yard, Doncaster, on hire, May 1999; returned to RFS; to Balfour Beatty, Leeds Station contract, on hire, about March 2000; returned to RFS, Doncaster, by 14th May 2001; to Blue Circle, Hope, Derbyshire, on hire, about February 2002; returned to Wabtec, November 2002; to GNER, Bounds Green, on hire, by May 2003; to Wabtec, by 27th July 2003; to Mendip Rail, Whatley Quarry, on hire, 5th January 2004; returned to Wabtec, 17th May 2004; to Daventry International Rail Freight Terminal, on hire, by 5th October 2004; to Wabtec, for repairs, 12th November 2004; to Daventry International Rail Freight Terminal, on hire, by 8th December 2004; to Hanson Quarry Products, Whatley Quarry, on hire, by 22nd January 2006; to Wabtec, Doncaster, by 11th June 2006; to GNER, Bounds Green, London, on hire, 16th February 2008.

D3765   Derby   1959   5A   11/86   P   08598/ HO16
08598   to Powell Duffryn Fuels Ltd, NCBOE Gwaun-cae-Gurwen, 16th January 1987; sold to RMS Locotec Ltd, Dewsbury, September 1995; to Celtic Energy, the new operators of Gwaun-cae-Gurwen, on hire; returned to RMS Locotec, Dewsbury, September 1998; to Cleveland Potash Ltd, Boulby, on hire, 21st June 1999; returned to RMS Locotec, Dewsbury, by 14th July 2000; sold to Potter Group, Knowsley, September 2001; to Potter Group, Selby, 24th May 2002; to Potter Group, Knowsley, by 12th April 2003; to Gloucestershire Warwickshire Railway, for gala, 7th July 2010; to Potter Group, Ely, week ending 3rd September 2010.

D3767   Derby   1959   30A   11/97   P   08600
08600   to A.V. Dawson Ltd, Middlesbrough (despatched from BR Eastleigh), 17th November 1997; to EWS Thornaby Depot, for wheel turning, 14th June 2004; returned to Dawson, 1st July 2004; to LH Group, Barton under Needwood, for repairs, 14th October 2008; returned to Dawson, 23rd March 2009.

D3769   Derby   1959   12A   3/86   P   AZ004
08602   to RFS (Engineering) Ltd, Kilnhurst, 30th June 1988; to Foster Yeoman Ltd, Isle of Grain, on hire, December 1988; to UK Paper, Kemsley, on hire, 1989; to Sheerness Steel Co Ltd, Sheerness, on hire, January 1990; to BREL Ltd, Litchurch Lane Carriage Works, Derby, on hire, March 1990; to BREL, York, on hire, November 1990; to BREL, Litchurch Lane, Derby, on hire, November 1990; purchased by Bombardier Transportation, Derby, 8th April 1991; to Fragonset, Derby, for overhaul, about September 2003; to Bombardier Transportation, Derby, about April 2004.

D3771   Derby   1959   16C   7/93   P   604/ PHANTOM
08604   to Great Western Society, Didcot, 28th September 1994.

D3780   Derby   1959   8J   12/93   P   08613
08613   to Trafford Park Estates Ltd, Manchester, 2nd February 1994; sold to Wabtec, Doncaster, November 2000; overhauled; to Trafford Park, on hire, 18th December 2000; sold to RT Rail, Crewe, January 2001; to Bombardier Transportation, Ilford, on hire, March 2001; to Wabtec, Doncaster, for overhaul, November 2006; returned to Ilford, on hire; to RMS Locotec, Wakefield, for repairs, 8th May 2007; to Barrow Hill Engine Shed Society, 5th October 2007; to Corus, Shotton, on hire, November 2007; to Castle Cement, Ketton, on hire, by 8th April 2009.

D3782   Derby   1959   8J   12/93   P   08615
08615   to Trafford Park Estates Ltd, Manchester, 2nd February 1994; sold to Wabtec, Doncaster, November 2000; to Hanson Quarry Products, Whatley Quarry, on hire, June 2002; to Merehead, on hire, October 2002; returned to Wabtec, 12th August 2003; to GNER, Craigentinny, on hire, by March 2004.

D3785   Derby   1959   52A   9/90   F   08618
08618   to Gwent Demolition, 1995; firm went bankrupt and sale cancelled; resold to T.J. Thomson Ltd, Stockton-on-Tees, for scrap, April 2000; dismantled at Gateshead Depot, for spares, April 2001; to T.J. Thomson Ltd, Stockton-on-Tees, rolling chassis only, April 2001; to Freightliner, Southampton, for spares, April 2001; scrapped on site (by Southampton Steel Ltd), September 2003.

D3789　　Derby　　　　　　　1959　　66B　　　12/95　P　08622/ HO28/19
08622　　to RMS Locotec, Dewsbury, 4th May 2002; to Faber Prest Ports, Flixborough Wharf, on hire, March 2003; to RMS Locotec, 19th April 2004; to PD Ports, Tees Dock, on hire, 6th August 2004; to RMS Locotec, Dewsbury, 14th February 2006; to RMS Locotec, Wakefield, 29th June 2006; to Faber Prest Ports, Flixborough Wharf, on hire, 9th October 2006; to Corus, Trostre, on hire, 3rd January 2008; to PD Ports, Tees Dock, on hire, March 2008; to Corus, Trostre, on hire, May 2009; to PD Ports, Tees Dock, on hire, week commencing 9th November 2009; to Weardale Railway, Wolsingham, for repairs, 18th January 2011; to Castle Cement, Ketton, on hire, 1st March 2011.

D3792　　Derby　　　　　　　1959　　2F　　　1/98　F　08625
08625　　to Dean Forest Railway (despatched from Canton Depot, Cardiff), 17th June 2000; to Cotswold Rail, Moreton in Marsh, about June 2001; to European Metal Recycling, Kingsbury, for scrap, February 2004; scrapped.

D3795　　Derby　　　　　　　1959　　2F　　　8/99　P　08628
08628　　to Goodman, Wishaw (despatched from Saltley), 28th September 2005; to European Metal Recycling, Kingsbury, about October 2005; to Goodman, Wishaw, 13th April 2006; to Bryn Engineering, and stored at Redrock Plant & Truck Services, Blackrod, Bolton, April 2006.

D3796　　Derby　　　　　　　1959　　ZN　　　6/95　P　08629/BRADWELL
08629　　to Railcare Ltd, Wolverton Works, June 1995; loco included in sale when works privatised; to Great Central Railway, for gala, 11th February 2011; to Alstom, Wolverton Works, 14th February 2011.

D3798　　Derby　　　　　　　1959　　31B　　12/92　P　08631/EAGLE
08631　　to Great Eastern Railway Company, County School Station, Norfolk, spring 1994; to Great Eastern Traction, Hardingham, Norfolk, 12th August 1995; to Fragonset, Derby, 13th December 1997; to Bombardier Transportation, Litchurch Lane Works, Derby, on hire, December 2003; returned to Fragonset (FM Rail), Derby, about June 2004; to Mid Norfolk Railway, Dereham, on hire, 22nd March 2007; returned to FM Rail, Derby; to Gwili Railway, on hire, 25th May 2010; to Mid Norfolk Railway, Dereham, 29th June 2011.

D3801　　Derby　　　　　　　1959　　30A　　2/93　F　08634
08634　　to Barrow Hill Engine Shed Society (HNRC), Staveley, 2nd July 2001; to West Coast Railway Company, Carnforth, 11th July 2002; scrapped, February 2005.

D3802　　Derby　　　　　　　1959　　81A　　2/04　P　08635
08635　　to T.J. Thomson, Stockton (despatched from EWS Toton Depot), 28th February 2007; to Severn Valley Railway, 26th April 2007.

D3810　　Horwich　　　　　　1959　　82B　　?　　P　08643
08643　　to Foster Yeoman Quarries, Merehead, 16th April 2003.

D3814　　Horwich　　　　　　1959　　36A　　4/97　F　08647
08647　　to South Yorkshire Railway Preservation Society, Meadowhall, Sheffield (despatched from Adtranz, Crewe), 12th March 1997; dismantled for spares, 7/8th November 1998; remains to Mayer Parry Ltd, Snailwell, Cambridgeshire, for scrap, 20th November 1998; scrapped.

D3815   Horwich   1959   ?   ?   P   OLD GEOFF/20
08648   purchased by R.T. Rail, Crewe (despatched from Laira Depot, Plymouth), moved direct to Wabtec, Doncaster, for repairs, 24th July 2002; to LNWR, Midland Road Depot, Leeds, for fitting with auto couplers, mid-2004; to Brunner Mond, Winnington Works, Northwich, on hire, about September 2004; to Wabtec, Doncaster, for repairs, about September 2005; to Brunner Mond, Northwich, on hire, 13th January 2006; to Wabtec, Doncaster, for repairs, August 2007; sold to RMS Locotec, Wakefield, November 2007; to PD Ports, Tees Dock, on hire, 17th January 2011.

D3816   Horwich   1959   ZG   6/95   P   08649/
08649   WOLVERTON
to Wessex Traincare Ltd, Eastleigh Works, June 1995; loco included in sale when works privatised; to Wimbledon Depot, for tyre turning, December 2000; returned to Eastleigh, January 2001; to LH Group, Barton under Needwood, for repairs, 14th October 2003; returned to Eastleigh, late 2003; to Alstom, Wolverton Works, 4th April 2006.

D3817   Horwich   1959   70D   8/89   P   08650
08650   to Foster Yeoman Quarries Ltd, Merehead Stone Terminal, Somerset, February 1989; to Foster Yeoman Quarries Ltd, Isle of Grain Works, Kent, 7th May 1989; to Foster Yeoman, Merehead, 28th June 2002; to Foster Yeoman, Isle of Grain, August 2002; to Wabtec, Doncaster, for repairs, about June 2003; to Foster Yeoman, Isle of Grain, late 2003; to Foster Yeoman, Merehead, for repairs, 25th February 2007; to Foster Yeoman, Isle of Grain, 2nd March 2007.

D3819   Horwich   1959   86A   7/92   P   08652
08652   to Foster Yeoman Quarries Ltd, Merehead Stone Terminal, Somerset, 5th June 1993; to ARC (Southern) Ltd, Whatley Quarry, Somerset, March 1995; to Merehead, 2001; to Whatley Quarry, 22nd October 2002; to Foster Yeoman Quarries, Acton Rail Terminal, London, 10th May 2004; to Whatley Quarry, October 2004; to Acton Rail Terminal, March 2005; to Whatley Quarry, June 2005; to Acton Rail Terminal, by 20th May 2007; to Whatley Quarry, by 9th September 2007; to LH Group, Barton under Needwood, for overhaul, March 2008; to Whatley Quarry, about October 2008.

D3822   Horwich   1959   40B   1/04   F   08655
08655   to T.J. Thomson Ltd, Stockton (despatched from EWS Thornaby Depot), 29th September 2005; to LH Group, Barton under Needwood, 16th June 2006; used for spares; remains scrapped by Donald Ward of Burton on Trent, April 2007.

D3832   Crewe   1960   40B   55C   P   08665
08665   to C.F. Booth Ltd, Rotherham, 16th December 2009; resold to HNRC (swapped for 08695); to Barrow Hill Engine Shed Society, Staveley, 17th December 2009.

D3835   Crewe   1960   5A   9/95   P   08668
08668   to South Yorkshire Railway Preservation Society (HNRC), Meadowhall, Sheffield (despatched from Adtranz, Crewe), 21st March 1997; to Barrow Hill Engine Shed Society (HNRC), Staveley, 3rd November 1999; to Wabtec, Doncaster, 2nd July 2001; some of its parts used in the rebuild of 08517; to HNRC, Long Marston, 20th December 2007; stripped for spares, February 2011; to European Metal Recycling, Kingsbury, for scrap, 29th June 2011; scrapped on arrival.

D3836 Crewe 1960 9A 5/89 P 08669
08669 to Trafford Park Estates Ltd, Manchester, 28th March 1989; sold to Wabtec, Doncaster, November 2000; to First Great Western, Laira Depot, Plymouth, on hire, 27th April 2001; to Wabtec, Doncaster, 14th November 2002; to Freightliner, Felixstowe, on hire, February 2003; to Wabtec, Doncaster, May 2003; to GNER, Bounds Green, on hire, 28th May 2003; to Wabtec, Doncaster, about March 2005.

D3837 Crewe 1960 66B 3/04 P 08670
08670 sold to T.J. Thomson Ltd, Stockton-on-Tees, January 2009; resold to Traditional Traction, Wishaw, Warwickshire, early 2009; resold to Colne Valley Enterprises Ltd, early 2009; to Colne Valley Railway, Essex (despatched from DBS Motherwell Depot), 5th March 2009.

D3845 Horwich 1959 41A 10/88 P 08678/ ARTILA
08678 to Glaxochem Ltd, Ulverston, Cumbria, 4th May 1989; to Steamtown, Carnforth, Lancashire, for repairs, 3rd August 1990; returned to Glaxochem; to Steamtown, Carnforth, for repairs, 20th July 1992; returned to Glaxochem; to Steamtown, Carnforth, 8th November 1994; to Fragonset, Derby, for repairs, May 2001; to Maintrain, Etches Park, on hire, by 9th June 2001; to Fragonset, by 3rd November 2001; to West Coast Railway Company, Carnforth, December 2001.

D3846 Horwich 1959 8J 6/76 F 08679
08679 to NCB North Gawber Colliery, Mapplewell, Barnsley, June 1976; to Royston Drift Mine, Barnsley, 7th September 1976; to North Gawber Colliery, 5th April 1979; to C.F. Booth Ltd, Rotherham, 18th April 1986; scrapped, April 1986.

D3849 Horwich 1959 ZF 6/95 P D3849/LIONHEART
08682 to ABB (Customer Support) Ltd, Doncaster Works, June 1995; loco included in sale when works privatised; to Bombardier Transportation, Derby, about January 2008; to Technical Centre, Derby, on hire, 31st July 2010; returned to Bombardier Transportation, Derby, October 2010.

D3850 Horwich 1959 16A 2/04 P 08683
08683 to Traditional Traction, Wishaw, Warwickshire, 8th March 2007; to Gloucestershire Warwickshire Railway, Toddington, for storage, November 2009; to Freightliner, Port of Felixstowe, on hire, 10th February 2011.

D3859 Horwich 1959 5A 1/94 F 692
08692 to ABB Transportation, Crewe, 5th May 1994; to ABB (Customer Support) Ltd, Doncaster, about August 1995; returned to ABB, Crewe, by 17th August 1996; sold to HNRC, 2002; to West Coast Railway Company, Carnforth, July 2002; scrapped, May 2005.

D3861 Horwich 1959 81A 2/04 P 08694
08694 to C.F. Booth Ltd, Rotherham, 8th January 2009; to Great Central Railway, Loughborough, 22nd May 2009.

D3862 Horwich 1959 66B 2/04 F 08695
08695 to T.J. Thomson Ltd, Stockton (despatched from EWS Ayr Depot), 27th July 2007; to Barrow Hill Engine Shed Society, Staveley, (HNRC), 11th September 2007; to C.F. Booth Ltd, Rotherham (swapped for 08665); scrapped December 2009.

D3864   Horwich   1959   16C   5/97   P   08697
08697   to Rail Vehicle Engineering Ltd, Derby (despatched from East Midlands Trains, Etches Park, Derby), July 2010.

D3866   Horwich   1960   5A   10/93   P   NPT
08699   to ABB, Crewe, 5th May 1994; sold to Cotswold Rail, Moreton in Marsh, early 2005; to Daventry International Rail Freight Terminal, on hire, about March 2005; to Cotswold Rail, May 2005; to Standard Gauge Steam Trust, Tyseley, for storage, 2005; to Allelys, Studley, Warwickshire, for storage, about 24th February 2006; sold to RMS Locotec, January 2007; to Corus, Shotton, for storage, 8th July 2009.

D3867   Horwich   1960   30A   9/93   P   D3867
08700   to Barrow Hill Engine Shed Society (HNRC), Staveley, 27th June 2001; to West Coast Railway Company, Carnforth, 7th August 2002; to Bryn Engineering Ltd, Wigan, 22nd June 2004; to Embsay & Bolton Abbey Railway, 23rd June 2004; to East Lancashire Railway, Bury, about April 2007.

D3871   Horwich   1960   1E   11/89   P   D3871
08704   despatched from BR Bletchley, 19th April 1990; to Port of Boston, Boston, 8th May 1990; to BR Doncaster Depot, for wheel turning, 20th January 1992; to Port of Boston, 29th January 1992; to Nene Valley Railway, Wansford, on loan, 3rd February 1993; to Port of Boston, 10th September 1997; to Wabtec, Doncaster, for repairs, 6th June 2001; returned to Port of Boston, 26th July 2001.

D3874   Crewe   1960   55G   8/93   F   08707
08707   to South Yorkshire Railway Preservation Society (HNRC), Meadowhall, Sheffield (despatched from Adtranz, Crewe), 4th April 1997; to Barrow Hill Engine Shed Society, Staveley, 27th July 2001; to West Coast Railway Company, Carnforth, 11th July 2002; scrapped, February 2005.

D3892   Crewe   1960   NTW   —-   P   08724
08724   to Great North Eastern Railway and allocated to Craigentinny Depot, Edinburgh; sold to RFS (Engineering) Ltd, Doncaster, 1997; to Foster Yeoman Ltd, Grain, Kent, on hire, 9th September 1998; to ARC, Whatley, on hire, October 1998; returned to RFS, Doncaster, 11th January 1999; to Blue Circle, Hope Cement Works, Derbyshire, on hire, March 1999; returned to RFS, Doncaster, 1999; to ARC, Whatley, on hire, by 19th September 1999; returned to RFS, Doncaster, 10th June 2000; to Northern Spirit, Neville Hill Depot, Leeds, on hire, about September 2000; to Wabtec, for repairs, early 2003; returned to Neville Hill Depot, on hire, by 19th June 2003.

D3896   Crewe   1960   66B   11/87   F   08728
08728   to Deanside Transit Ltd, Glasgow, November 1987; to HNRC, Long Marston, 19th June 2007; to C.F. Booth Ltd, Rotherham, for scrap, 4th November 2009; scrapped, November 2009.

D3898   Crewe   1960   ZH   6/95   P   08730/ THE CALEY
08730   to Railcare Ltd, Springburn Works, Glasgow, June 1995; loco included in sale when works privatised; to LH Group, Barton under Needwood, for repairs, 19th August 2009; returned to Glasgow, about March 2010.

D3899　　Crewe　　　　　　1960　　66B　　12/95　F　08731
08731　　went to Swindon Works for dual brake fitting, June 1983; found to have frame damage so identity changed with 08572; although in reality it was 08572, it emerged from Swindon Works numbered 08731; to T.J. Thomson Ltd, Stockton (despatched from EWS Motherwell Depot), 21st June 2002; to Foster Yeoman Quarries, Merehead, 13th August 2003; to LH Group, Barton under Needwood, for stripping, about March 2004; frame and wheels returned to Merehead, 16th April 2004; remains to Bodmin & Wenford Railway, for cannibalising, August 2008; remains to Merehead, 29th August 2008; to J.W. Ransome & Sons, Frome, for scrap, March 2009; scrapped March 2009.

D3902　　Crewe　　　　　　1960　　2F　　　8/96　　P　08734
08734　　to Dean Forest Railway, Lydney, (despatched from Canton Depot, Cardiff), 17th June 2000.

D3904　　Crewe　　　　　　1960　　66B　　11/87　F　4/ 08736
08736　　to Deanside Transit Ltd, Glasgow, November 1987; to HNRC, Long Marston, 23rd May 2007; to C.F. Booth Ltd, Rotherham, for scrap, 5th November 2009; scrapped, November 2009.

D3908　　Crewe　　　　　　1960　　?　　　9/97　　F　08740
08740　　to T.J. Thomson Ltd, Stockton (despatched from Ferrybridge), about October 2005; to LH Group, Barton under Needwood, 15th June 2006; used for spares; remains scrapped by Donald Ward of Burton on Trent, April 2007.

D3911　　Crewe　　　　　　1960　　55H　　3/93　P　BRYAN TURNER
08743　　to RFS (Engineering) Ltd, Doncaster, March 1993; to Grovehurst UK Paper Ltd, Sittingbourne, on hire, 25th April 1993; returned to RFS, Doncaster, 1996; sold to ICI, Billingham, 27th January 1997; to ICI, Wilton, April 2004; to Cleveland Potash Ltd, Teesport, on hire, about July 2005; returned off hire, October 2005.

D3914　　Crewe　　　　　　1960　　66B　　2/99　　F　08746
08746　　to Barrow Hill Engine Shed Society, Staveley (despatched from Doncaster), 21st July 2003; used for spares to repair 08928; remains to C.F. Booth Ltd, Rotherham, for scrap, 31st August 2003; scrapped, January 2004.

D3918　　Crewe　　　　　　1960　　81A　　6/98　　P　08750
08750　　sold to R.T. Rail, Crewe, 2000; to Wabtec Rail, Doncaster (despatched from Stratford Depot), for assessment, 7th August 2000; to Wessex Traincare Ltd, Eastleigh Works, on hire, 1st December 2000; returned to R.T. Rail; to Ilford Depot, on hire, 29th January 2001; to Wabtec, Doncaster, for repairs, by 18th November 2001; to Channel Tunnel Rail Link, Beechbrook Farm, near Ashford, on hire, by March 2002; to Wabtec, Doncaster, 1st August 2002; to Wensleydale Railway, 21st July 2003; to Wabtec, December 2003; to Imreys Minerals, Quidhampton, Salisbury, on hire, 26th March 2004; to Wabtec, for repairs, about April 2005; returned to Quidhampton, about August 2005; to Wabtec, for repairs, 13th October 2005; to Tubelines, Ruislip, London, on hire, 16th May 2006; to First Capital Connect, Hornsey Depot, on hire, 30th May 2007; to RMS Locotec, Wakefield, for repairs, about November 2007; returned to Hornsey Depot; to Bombardier Transportation, Ilford Depot, on hire, about April 2011.

D3919　　Crewe　　　　　　1960　　41A　　5/97　　F　08751
08751　　to RFS Engineering, Doncaster, by June 1998; used for spares; remains to C.F. Booth Ltd, Rotherham, January 2004; re-sold to R.T. Rail; to Wabtec, Doncaster,

for spares recovery, April 2004; to C.F. Booth Ltd, Rotherham, for scrap, 7th July 2004; scrapped, 4th September 2004.

D3922 Horwich 1961 60A 8/99 P 08754/ HO41
08754 to R.T. Rail, Crewe, August 1999; to Wabtec, Doncaster, autumn 1999; to Port of Felixstowe, on hire, about April 2000; to Freightliner, Garston, on hire, 2000; to Freightliner, Dagenham Dock, on hire, by 24th November 2001; to Wabtec, Doncaster, for repairs, by 16th February 2003; to Grant Rail, March, on hire, 22nd April 2003; to Silverlink, Bletchley, on hire, 31st March 2004; to Reading PW Yard, by November 2004; sold to RMS Locotec, Dewsbury, about May 2005; to PD Ports, Tees Dock, on hire, by 27th June 2005; to RMS Locotec, 1st September 2005; to PD Ports, Tees Dock, on hire, 15th September 2005; to RMS Locotec, 25th November 2005; to Network Rail, Whitemoor Yard, March, on hire, 31st March 2006; to Bombardier Transportation, Ilford, on hire, April 2008; to Network Rail, Whitemoor Yard, on hire, by July 2008; to Wabtec, Doncaster, October 2008.

D3924 Horwich 1961 86A ? P 08756
08756 to R.T. Rail, Crewe (despatched from Canton Depot, Cardiff), 9th September 2003; to Wabtec, Doncaster, 9th September 2003; to Bombardier, Doncaster, for storage, April 2004; to West Yard, Doncaster, for storage, July 2005; to Wabtec, Doncaster, for overhaul, 2006; to Brunner Mond, Northwich, on hire, September 2006; to RMS Locotec, on hire, 1st November 2006; used on Stirling to Alloa line contract; to Elsecar Steam Railway, near Barnsley, 14th November 2006; to RMS Locotec, Wakefield, about November 2007; to Network Rail, Whitemoor Yard, March, on hire, 2nd October 2008; to Corus, Shotton Steelworks, on hire, 6th April 2009.

D3930 Horwich 1961 60A 8/99 P OLD TOM
08762 to R.T. Rail, Crewe, August 1999; to Wabtec, Doncaster, September 1999; to Freightliner, Dagenham Dock, on hire, by 26th April 2000; to Port of Felixstowe, on hire, after 13th December 2001; to Channel Tunnel Rail Link, Beechbrook Farm, near Ashford, on hire, by February 2002; to Europort, Wakefield, on hire, May 2002; to Wabtec, Doncaster, for repairs, by 10th June 2002; to Freightliner, Dagenham, on hire, about September 2002; to Imreys Minerals (ECC), Quidhampton, on hire, 16th January 2003; to Wabtec, Doncaster, for repairs, 30th March 2004; to Freightliner, Midland Road Depot, Leeds, on hire, by 28th July 2004; fitted with auto couplers, 2004; to Brunner Mond, Winnington Works, Northwich, on hire, about September 2004; suffered collision damage, 2006; to Wabtec, for repairs, 19th April 2006; returned to Brunner Mond, on hire; sold to Wabtec, Doncaster, August 2007.

D3932 Horwich 1961 64B 5/88 P 003/ FLORENCE
08764 to RFS (Engineering) Ltd, Kilnhurst, May 1988; to ARC Ltd, Machen Quarry, Gwent, on hire, 26th September 1990; returned to RFS, 31st October 1990; to BREL Ltd, York, on hire, 1st November 1990; to Flixborough Wharf Ltd, Scunthorpe, on hire, 13th December 1990; to RFS, Doncaster, 8th February 1993; to Channel Tunnel (number 97), on hire, 10th February 1993; to RFS, Doncaster, June 1993; to Sheerness Steel Co, on hire, June 1993; returned to RFS, 1993; to Hartlepool Power Station, on hire, October 1993; returned to RFS, Doncaster, by 10th April 1994; to Flixborough Wharf Ltd, Scunthorpe, on hire, June 1994; returned to RFS, Doncaster, by 5th December 1996; to Transfesa Rail Terminal, Tilbury, on hire, 21st August 1997; sold to Transfesa, September 1997; to RFS, Doncaster, for repairs, by February 1998; to Transfesa Rail Terminal, Tilbury, by 13th January 1999.

D3933   Horwich   1961   81A   5/08   P   08765
08765   to HNRC (despatched from DBS Eastleigh); to Boden Rail Engineering, Washwood Heath, 28th June 2011.

D3935   Horwich   1961   30A   1/94   P   D3935
08767   to North Norfolk Railway, Sheringham (despatched from Colchester Depot), 22nd August 1994.

D3937   Derby   1960   87E   5/89   P   D3937/ GLADYS
08769   to MoD Long Marston, Worcestershire, 20th April 1990; to The Fire Service College, Moreton in Marsh, 12th November 1991; to South Wales Diesel Locomotive Association, Dean Forest Railway, 2nd March 2000; to Severn Valley Railway, 12th May 2003.

D3940   Derby   1960   30A   1/94   P   D3940
08772   to East Anglian Railway Museum, Essex (despatched from Colchester), by 27th March 1994; to North Norfolk Railway, Sheringham, 18th September 2001.

D3941   Derby   1960   16A   3/94   P   D3941
08773   sold to Mike Darnall, Newton Heath, Manchester, 25th July 2000; to Embsay & Bolton Abbey Railway, February 2006.

D3942   Derby   1960   51L   10/88   P   ARTHUR VERNON
08774   DAWSON
to A.V. Dawson Ltd, Middlesbrough, September 1988; to Cobra Railfreight, Middlesbrough, on hire, October 1998; returned to Dawson, 1998; to Wabtec, Doncaster, for overhaul, late 2001; returned to Dawson, 22nd February 2002.

D3948   Derby   1960   ?   ?   P   08780/FRED
08780   to Cotswold Rail, Moreton in Marsh (despatched from Landore Depot), about June 2001; to East Lancashire Railway, Bury, November 2002; to Transplant Ltd (London Underground maintenance), West Ruislip, on hire, 26th April 2005; to LNWR Ltd, Midland Road Depot, Leeds, 15th September 2005; to Wabtec, Doncaster, 27th March 2006; to LNWR Ltd, Crewe, 1st June 2006; to Southall MPD, London, 21st December 2007.

D3953   Derby   1960   86A   3/89   P   004/ CLARENCE
08785   to RFS (Engineering) Ltd, Kilnhurst, 25th September 1990; to Grovehurst Energy, Sittingbourne, Kent, on hire, 4th June 1991; to Channel Tunnel Rail Link, on hire, 10th September 1992; to RFS, Doncaster, July 1993; to BASF Chemicals Ltd, Seal Sands, on hire, 9th August 1994; returned to RFS, Doncaster, 25th November 1994; sold to Freightliner, 14th July 1997.

D3954   Derby   1960   16A   12/08   P   08786
08786   to HNRC; to Barrow Hill Engine Shed Society, Staveley (despatched from DBS Doncaster Depot), 24th January 2011.

D3956   Derby   1960   16C   1/94   P   08788
08788   to Great Central Railway, Loughborough, 30th March 1994; to R.T. Rail, Crewe, 24th February 1999; to Scot Rail, Inverness, on hire, 1st September 1999; to Manchester Ship Canal Company, Barton Dock, on hire, 21st April 2005; to Wabtec, Doncaster, for repairs, 2nd June 2005; to Scot Rail, Inverness, on hire, 16th November 2005; sold to RMS Locotec, Wakefield, about November 2007.

D3969   Derby   1960   86A   6/00   F   08801
08801   to R.T. Rail, Crewe, 8th September 2003; to Wabtec, Doncaster, 8th September 2003; to C.F. Booth Ltd, Rotherham, for scrap, February 2004; scrapped, 24th March 2004.

D3975   Derby   1960   66B   6/04   P   08807
08807   to T.J. Thomson Ltd, Stockton, 26th April 2007; to EWS Thornaby Depot, for wheel turning, 30th October 2007; to A.V. Dawson Ltd, Middlesbrough, for spares, week ending 23rd November 2007.

D3977   Derby   1960   8J   12/93   P   08809
08809   to Otis Euro Trans Rail Ltd, Salford, Manchester, December 1993; to Flixborough Wharf Ltd, Scunthorpe, on hire, week ending 12th January 1996; returned to Otis; sold to Harry Needle Railroad Company, late 1999; to Fragonset, Derby, for certification, about March 2000; to Barrow Hill Engine Shed Society (HNRC), Staveley, 13th June 2000; to Freightliner, Coatbridge, on hire, 29th June 2000; to Motherwell shed, for repairs, September 2001; returned to Freightliner, Coatbridge, on hire; to Barrow Hill Engine Shed Society, Staveley, 24th May 2002; sold to Cotswold Rail, Moreton in Marsh, 29th June 2002; to Anglia Railways, Crown Point Depot, Norwich, on hire, 24th October 2002; to Ilford Depot, for tyre turning, 10th December 2003; returned Crown Point, 17th December 2003; to Brush, Loughborough, for overhaul, 17th March 2005; to Allelys, Studley, Warwickshire, for storage, 7th February 2006; sold to RMS Locotec, Wakefield, October 2007; to Corus, Shotton, for storage, 26th October 2009; to Boden Rail Engineering, Washwood Heath, 29th November 2010.

D3978   Derby   1960   NC   ?   P   08810
08810   to Cotswold Rail, Moreton in Marsh, about June 2001; to Brush, Loughborough, for overhaul, 15th August 2001; to Anglia Railways, Crown Point Depot, Norwich, on hire, 18th September 2001; to Railway Age, Crewe, 10th December 2003; to LNWR Ltd, Carriage Works, Crewe, about March 2004.

D3981   Derby   1960   51L   1/00   F   08813
08813   to HNRC, Long Marston (despatched from EWS Thornaby Depot), 26th September 2006; stripped for spares, January 2011; remains to T.J. Thomson Ltd, Stockton, for scrap, 4th February 2011; scrapped, 9th February 2011.

D3984   Derby   1960   51L   2/86   F   HO25
08816   to Cobra Railfreight Ltd, Middlesbrough, 15th February 1986; to Harry Needle Railroad Company, 1995; to Johnson Ltd, Widdrington Disposal Point, on hire, 19th August 1995; to RFS (Engineering) Ltd, Doncaster, 10th July 1998; scrapped, August 1999.

D3986   Derby   1960   5A   4/97   P   08818/ MOLLY
08818   to Railway Age, Crewe, September 1997; sold to Harry Needle Railroad Company, March 1997; to LNWR Ltd, Carriage Works, Crewe, on hire, by June 1998; to Port of Felixstowe, on hire, 13th August 1999; to Barrow Hill Engine Shed Society (HNRC), Staveley, 3rd November 1999; to Freightliner, Basford Hall, Crewe, on hire, 10th June 2000; to Barrow Hill Engine Shed Society, 12th April 2001; to Freightliner, Stourton, Leeds, on hire, 24th January 2002; to Battlefield Line, Shackerstone, 21st August 2002; to Freightliner, Coatbridge, on hire, 4th November 2002; to Battlefield Line, Shackerstone, 4th December 2002; to Freightliner, Calvert Landfill Site, on hire, 12th May 2003; to MoD Bicester, June 2004; to Severn Valley Railway, on hire, 21st

September 2004; to Network Rail, Whitemoor Yard, March, on hire, about October 2004; to HNRC, Long Marston,12th October 2005; to Barrow Hill Engine Shed Society, Staveley, 21st March 2006; to Flixborough Wharf, Scunthorpe, on hire, 20th May 2007.

D3987   Derby   1960   86A   9/99   F   08819
08819   to R.T. Rail, Crewe, September 2003; to Wabtec, Doncaster, for storage, about November 2003; to Bombardier, Doncaster, for storage, April 2004; to West Yard, Doncaster, for storage, by July 2005; to C.F. Booth Ltd, Rotherham, for scrap, 21st May 2008; scrapped, August 2008.

D3991   Derby   1960   ZF   6/95   P   08823/LIBBIE
08823   to ABB (Customer Support) Ltd, Doncaster Works, June 1995; loco included in sale when works privatised; to Churnet Valley Railway, Cheddleton, Staffordshire, November 2000; to LH Group, Barton under Needwood, for repairs, 3rd August 2007; purchased by Hunslet Engine Company, March 2008; to Manchester Ship Canal Company, Trafford Park, on hire, 24th March 2008; to LH Group, Barton under Needwood, for repairs, 22nd May 2009; to Thames Steel, Sheerness, on hire, 8th June 2010; to LH Group, Barton under Needwood, for repairs, 13th August 2010; to Thames Steel, Sheerness, on hire, 20th August 2010.

D3993   Derby   1960   81A   12/99   P   08825
08825   to Battlefield Line, Shackerstone (despatched from EWS Wigan Depot), 7th October 2005.

D3994   Derby   1960   66B   12/95   F   08826
08826   to T.J. Thomson Ltd, Stockton (despatched from EWS Motherwell Depot), July 2002; to East Somerset Railway, Cranmore, 8th October 2003; to Foster Yeoman, Merehead, 9th June 2004; to Mid Hants Railway, Ropley, for storage, March 2009; to Knight's Rail Services, Eastleigh Works, 21st October 2010; used for spares; remains scrapped, April 2011.

D3995   Derby   1960   66B   3/00   P   08827
08827   to Barrow Hill Engine Shed Society (HNRC), Staveley (despatched from EWS Motherwell Depot), 2nd September 2005; to HNRC, Long Marston, week ending 28th July 2006; stripped for spares, January 2011; to European Metal Recycling, Kingsbury, for scrap, 7th July 2011.

D3997   Derby   1960   16A   6/93   F   08829
08829   to European Metal Recycling, Kingsbury, August 2000; to Barrow Hill Engine Shed Society, Staveley, 19th March 2001; to West Coast Railway Company, Carnforth, 7th October 2002; scrapped, February 2005.

D3998   Derby   1960   NTW   —-   P   08830
08830   to East Somerset Railway, Cranmore (on lease from Cardiff Railways Ltd and despatched from Cardiff Cathays), 2nd October 1996; to Foster Yeoman Quarries Ltd, Torr Works, on hire, July 1997; returned to ESR, about August 1997; to Cardiff Railways Ltd, off lease, 18th September 1999; to LNWR Ltd, Carriage Works, Crewe, on hire, 30th December 1999; to Crewe Works, for open day, 19th May 2000; returned to LNWR, Crewe, May 2000; to Freightliner, Felixstowe, on hire, about 12th April 2005; to Heritage Centre, Crewe, 13th July 2005; to LNWR, Midland Road Depot, Leeds, on hire, by 16th August 2005; to Wabtec, Doncaster, 2007; to Heritage Centre, Crewe, 2007.

D4002   Derby   1960   NTW   —-   P   08834
08834   to Transmanche-Link, Dolland Moor, Kent, on hire, 23rd September 1992; sold to RFS (Engineering) Ltd, Doncaster, 27th April 1993; to GNER, Bounds Green, on hire, 1997; returned to RFS, Doncaster, by 14th February 1998; to GNER, Bounds Green, on hire, 2nd August 1999; to Wabtec, Doncaster, 28th May 2003; to Foster Yeoman Quarries, Merehead, on hire, 19th January 2004; to Daventry International Rail Freight Terminal, on hire, November 2004; to Wabtec, Doncaster, 26th May 2005; to Foster Yeoman Quarries, Merehead, on hire, early 2006; returned to Wabtec, about April 2006; sold to Direct Rail Services, May 2006; to DRS, Carlisle, 2nd October 2006; to DRS, Gresty Lane, Crewe, 23rd January 2007; sold to HNRC, January 2009; to Basford Hall, Crewe, 22nd May 2009; to SERCO, Old Dalby, 27th May 2009.

D4014   Horwich   1961   8J   9/89   P   003
08846   to BREL Ltd, Litchurch Lane Carriage Works, Derby, October 1989; to ABB Transportation, Crewe, October 1993; to ABB Transportation, York, September 1995; to ABB, Crewe, for overhaul, 30th August 1996; to Adtranz, Derby, 15th May 1998; to Fragonset, Derby, for overhaul, about March 2004; to Bombardier Transportation, Litchurch Lane Works, Derby, about June 2004.

D4015   Horwich   1961   ZG   6/95   P   08847
08847   to Wessex Traincare Ltd, Eastleigh Works, June 1995; loco included in sale when works privatised; sold to Cotswold Rail, Moreton in Marsh, 18th May 2001; to Brush, Loughborough, for overhaul, 24th July 2001; to Cotswold Rail, about July 2002; to Anglia Railways, Norwich, on hire, September 2002; to GBRf, Europort, Doncaster, on hire, November 2002; to Wabtec, Doncaster, 28th March 2003; to British Gypsum, Mountfield, East Sussex, on hire, 7th May 2003; to Anglia Railways, Norwich, on hire, 11th September 2003; to Horton Road Depot, Gloucester, for storage, 2006; purchased by RMS Locotec, Wakefield, October 2007; to Wabtec, Doncaster, early 2008; to Anglia Railways, Norwich, on hire, by 26th March 2008.

D4018   Horwich   1961   81D   12/92 P   4018
08850   to West Somerset Railway, Minehead, 13th September 1993; to North Yorkshire Moors Railway, 11th March 1998.

D4021   Horwich   1961   EC   1/96   P   08853
08853   to RFS (Engineering) Ltd, Doncaster, 30th January 1997; overhauled; to Great North Eastern Railway, Bounds Green, on hire, about March 1997; returned to RFS, Doncaster, 1998; to GNER, Bounds Green, on hire, June 2003; to Wabtec, Doncaster, 8th February 2008.

D4035   Darlington   1960   8F   ?   F   HL1007
08867   to RMS Locotec Ltd, Dewsbury, about 1993; to Brunner Mond, Northwich, on hire, 6th January 1994; to RMS Locotec, off hire, by March 1998; to Brunner Mond, on hire, 1998; to RMS Locotec, Dewsbury, for repairs, 29th November 1999; to Marcroft, Horbury, Wakefield, on hire, early 2000; returned to RMS Locotec, by February 2001; to Cobra Railfreight, Wakefield, on hire, 2nd July 2002; to Ferrybridge Depot, for storage, September 2002; to T.J. Thomson Ltd, Stockton, for scrap, 4th October 2005; scrapped, 20th June 2007.

D4036   Darlington   1960   31B   12/92 P   08868
08868   to South Yorkshire Railway Preservation Society (HNRC), Meadowhall,

Sheffield, 22nd February 1994; to East Lancashire Railway, Bury, 16th April 1994; to RFS (Engineering) Ltd, Doncaster, on hire, 3rd September 1997; sub-hired to Fastline Track Renewals, Peterborough, 12th September 1997; used at May-Gurney, Connington Tip, near Peterborough; to Railway Age, Crewe, 2nd June 1998; to Freightliner, Basford Hall, Crewe, on hire, January 1999; to MoD, Long Marston, on hire, 2nd February 1999; to Railway Age, Crewe, 6th April 1999; to Port of Felixstowe, on hire, 6th August 1999; to LNWR Ltd, Carriage Works, Crewe, on hire, 13th August 1999; to Freightliner, Basford Hall, Crewe, on hire, 18th September 2000; to Freightliner, Trafford Park, on hire, 30th April 2001; to Barrow Hill Engine Shed Society, Staveley, 17th April 2002; to Blue Circle, Hope Cement Works, Derbyshire, on hire, 15th November 2002; to Port of Felixstowe, on hire, 25th July 2003; to Freightliner, Midland Road Depot, Leeds, on hire, 7th December 2003; to Port of Felixstowe, on hire, 28th July 2004; to LNWR Ltd, Carriage Works, Crewe, 24th November 2004.

D4037 Darlington 1960 NC ? F 08869
08869 sold to Cotswold Rail, Moreton in Marsh, about June 2001; to Mid Norfolk Railway, Dereham (despatched from Crown Point Depot, Norwich), for road transfer, 14th August 2001; to Brush, Loughborough, for repairs, 15th August 2001; to Barrow Hill Engine Shed Society (HNRC), Staveley, 27th August 2003; to HNRC, Long Marston, for storage, September 2006; to European Metal Recycling, Kingsbury, for scrap, 29th September 2010; scrapped, January 2011.

D4038 Darlington 1960 55G 5/93 P HO24
08870 to South Yorkshire Railway Preservation Society, Meadowhall, Sheffield, 1st March 1994; to Cobra Railfreight, Wakefield, on hire, 9th December 1994; to SYRPS, Sheffield, 15th October 1997; sold to RMS Locotec Ltd, Dewsbury, September 1998; to Anglian Railways, Norwich, on hire, 26th September 1998; returned to RMS Locotec, for repairs, 30th March 1999; to Anglian Railways, Norwich, on hire, February 2000; to RMS Locotec, for repairs, 1st March 2001; to Anglian Railways, Norwich, on hire, about March 2001; to RMS Locotec, 14th June 2001; to Bombardier Transportation, Horbury, Wakefield, on hire, 19th June 2001; to Redland, Barrow upon Soar, on hire, by 12th January 2002; to Ford, Bridgend, on hire, March 2002; to Redland, Barrow upon Soar, on hire, 6th November 2002; to RMS, 13th October 2004; to Castle Cement, Ketton, on hire, 27th January 2005; to Weardale Railway, Wolsingham, for repairs, 6th July 2010.

D4039 Darlington 1960 41A 10/90 P 08871
08871 to Humberside Sea & Land Services, Grimsby (despatched from BR Immingham), 16th December 1990; sold to Cotswold Rail, Moreton in Marsh, by 7th April 2001; to Brush, Loughborough, for repairs, 18th April 2001; to Anglia Railways, Norwich, on hire, August 2001; to Wabtec, Doncaster, for repairs, 5th November 2002; to Bombardier, Ilford, for tyre turning, January 2004; to Network Rail, Whitemoor Yard, March, on hire, 1st April 2004; to Daventry International Rail Freight Terminal, on hire, 8th October 2004; to Brush, Loughborough, for repairs, 9th November 2004; returned to Daventry International Rail Freight Terminal, 14th December 2004; to Brush, for repairs, 11th April 2005; to Anglia Railways, Crown Point Depot, Norwich, on hire, 28th November 2005; to Horton Road Depot, Gloucester, for storage, about August 2006; purchased by RMS Locotec, Wakefield, October 2007; to Anglia Railways, Crown Point Depot, Norwich, on hire, by 26th May 2007; to Wabtec, Doncaster, for repairs, by August 2008; to East Coast Trains, Craigentinny, on hire, 27th November 2010.

D4040 Darlington 1960 40B 2/04 P 08872
08872 to European Metal Recycling, Sheffield (despatched from DBS, Immingham Depot), 20th August 2010.

D4041 Darlington 1960 5A ? P 08873
08873 to ABB Transportation, Derby, 15th May 1998; sold to R.T. Rail, Crewe, about April 2000; to RFS (Engineering) Ltd, Doncaster, for repairs, May 2000; to LNWR Ltd, Carriage Works, Crewe, on hire, by 7th September 2000; to Manchester Ship Canal Company, Barton Dock, on hire, May 2005; sold to LH Group, Barton under Needwood, November 2005 (remained on hire to MSC Barton Dock); to LH Group, for repairs, 20th June 2006; to Manchester Ship Canal Company, Barton Dock, on hire, 14th November 2006; to LH Group, for repairs, 24th April 2007; to Port of Felixstowe, on hire, 22nd October 2007; to Manchester Ship Canal Company, Barton Dock, on hire, 7th January 2009; to Freightliner, Trafford Park, on hire, May 2009; to LH Group, Barton under Needwood, for repairs, 20th August 2009; to Innovative Logistics, Brierley Hill, on hire, November 2009; to Freightliner, Southampton, on hire, 20th May 2010; to LH Group, for repairs, 15th February 2011; to Freightliner, Southampton, on hire, week commencing 4th July 2011.

D4042 Darlington 1960 55H 2/92 P 08874
08874 to RFS (Engineering) Ltd, Kilnhurst; 2nd July 1992; to Thames Waste Management (number 97), on hire, 13th February 1993; returned to RFS, Doncaster, 11th February 1994; to Teesbulk Handling, Middlesbrough, on hire, 2nd June 1994; returned to RFS, Doncaster, 13th September 1994; to Sheerness Steel Co Ltd, Sheerness, on hire, 16th September 1995; returned to RFS, Doncaster, May 1997; purchased by R.T. Rail, Crewe, October 1998; to Crewe for certification; to Hays Chemicals, Sandbach, on hire, 12th October 1998; to RFS (Engineering) Ltd, Doncaster, for repairs, 8th January 1999; to Silverlink, Bletchley, on hire, 26th February 1999; to LH Group, Barton under Needwood, for TPWS fitment, early April 2004; to Silverlink, Bletchley, on hire, August 2004; sold to RMS Locotec, Wakefield, about November 2007; to Dartmoor Rail, Meldon Quarry, on hire, 19th December 2007; to Mid Norfolk Railway, Dereham, 14th April 2008; to Anglia Railways, Crown Point Depot, Norwich, on hire, late April 2008.

D4043 Darlington 1960 51L 5/91 F 08875
08875 to RFS (Engineering) Ltd, Kilnhurst, 20th November 1991; used for spares; scrapped on site by C.F. Booth Ltd, Rotherham, August 1993.

D4044 Darlington 1960 36A 10/91 F 08876
08876 to RFS (Engineering) Ltd, Kilnhurst (despatched from BR Tinsley), 2nd July 1992; to RFS, Doncaster, autumn 1993; used for spares; remains scrapped on site by Hudsons of Telford, April 1994.

D4056 Darlington 1961 40B 6/72 F D4056/ 55
to NCB Ashington Central Workshops, January 1973; to Shilbottle Colliery, February 1973; to Ashington Central Workshops, March 1974; to Shilbottle Colliery, 18th June 1974; scrapped on site by T.J. Thomson Ltd of Stockton-on-Tees, March 1983.

D4067 Darlington 1961 41J 12/70 P 10119
to NCB Betteshanger Colliery, Kent, April 1971; to Snowdown Colliery, Kent, 27th May 1976; to Nailstone Colliery, Leicestershire, 14th June 1976; to BR Doncaster, for repairs, 28th October 1976; to Nailstone Colliery, 23rd December 1976; to Great Central Railway, Loughborough, 6th February 1980.

D4068 Darlington 1961 40B 6/72 F No.56/ 9300-116
to NCB Ashington Central Workshops, January 1973; to Shilbottle Colliery, 16th February 1973; to Lambton Engine Works, Philadelphia, 3rd April 1979; to Whittle Colliery, 25th April 1980; scrapped on site by C.F. Booth Ltd of Rotherham, October 1985.

D4069 Darlington 1961 41J 4/72 F 9300-111
to NCB Ashington Central Workshops, September 1972; to Whittle Colliery, 20th October 1972; to Lambton Engine Works, Philadelphia, 28th February 1978; to Whittle Colliery, 30th March 1979; to C.F. Booth Ltd, Rotherham, November 1985; scrapped, December 1985.

D4070 Darlington 1961 41J 4/72 F No.52/ 9300-112
to NCB Ashington Central Workshops, September 1972; to Shilbottle Colliery, 9th February 1973; to Ashington Central Workshops, 23rd September 1974; to Lambton Engine Works, Philadelphia, for rebuild, 11th October 1975; to Bates Colliery, Blyth, 12th November 1976; to Whittle Colliery, 16th April 1977; to Lambton Engine Works, 29th April 1980; to Whittle Colliery, 5th June 1981; scrapped on site by C.F. Booth Ltd of Rotherham, October 1985.

D4072 Darlington 1961 31B 4/72 F No.53/ 9300-114
to NCB Ashington Central Workshops, October 1972; to Whittle Colliery, 1st November 1972; to Lambton Engine Works, Philadelphia, 14th June 1977; to Whittle Colliery, 28th February 1978; to Lambton Engine Works, 22nd June 1978; to Whittle Colliery, 16th August 1978; to Lambton Engine Works, 29th September 1978; to Whittle Colliery, 26th November 1978; to Lambton Engine Works, 19th December 1979; to Whittle Colliery, 5th June 1980; to Lambton Engine Works, 14th April 1981; to South Hetton Colliery, 10th May 1982; to Lambton Engine Works, 27th September 1982; to Ashington Colliery, July 1983; scrapped on site by C.F. Booth Ltd of Rotherham, November 1985.

D4074 Darlington 1961 31B 4/72 F No.54
to NCB Ashington Central Workshops, October 1972; to Whittle Colliery, 15th December 1972; to Lambton Engine Works, Philadelphia, 8th February 1977; scrapped, August 1978.

D4092 Darlington 1960 34E 9/68 P D4092
Powell Duffryn Fuels Ltd, NCBOE Gwaun-cae-Gurwen Disposal Point, Glamorgan, October 1968; to BR Canton Depot, Cardiff, for repairs, August 1977; returned to Gwaun-cae-Gurwen, October 1977; to South Yorkshire Railway Preservation Society (HNRC), Meadowhall, Sheffield, 26th October 1988; to Barrow Hill Engine Shed Society (HNRC), Staveley, 26th July 2001.

D4095 Horwich 1961 66B 2/04 P 881
08881 to Traditional Traction, Wishaw, Warwickshire, 20th April 2007; to Alstom, Wembley, for wheel turning, 15th August 2007; returned to Wishaw, 19th September 2007; to Gloucestershire Warwickshire Railway, Toddington, for storage, early 2008; to Lafarge Cement, Barrow-on-Soar, on hire, 5th October 2008; to Gloucestershire Warwickshire Railway, Toddington, December 2008.

D4115 Horwich 1962 36A 5/93 P 08885/18/HO42
08885 to Great Central Railway, Ruddington, Nottingham, 15th June 1994; to Midland Railway, Butterley, 7th November 2004; purchased by R.T. Rail, Crewe and

sent to RMS Locotec, Dewsbury for overhaul, 25th August 2005; to Network Rail, Whitemoor Yard, March, on hire, 31st October 2005; to PD Ports, Tees Dock, on hire, 3rd April 2009.

D4122   Horwich   1962   70D   12/96   P   08892
08892   to RFS (Engineering) Ltd, Doncaster, 11th December 1996; to GNER, Bounds Green, on hire, 18th April 1997; returned to RFS, Doncaster, 2nd August 1999; to GNER, Bounds Green, on hire, by 21st April 2001; to Wabtec, Doncaster, 10th April 2003; to Soho EMU Depot, Birmingham, on hire, October 2004; to Central Trains, Tyseley, on hire, 12th November 2004; returned to Wabtec, June 2005; sold to Direct Rail Services, May 2006; to DRS, Kingmoor, Carlisle, 25th July 2006; to DRS, Gresty Lane Depot, Crewe, about January 2007; to North Pole Depot, on hire, October 2007; to DRS, Gresty Lane Depot, Crewe, 14th November 2007; to Fastline, Doncaster, on hire, early July 2008; sold to HNRC, about August 2008; to Lafarge Cement, Hope, on hire, by 6th September 2008; to Bombardier Transportation, Derby, on hire, 14th October 2010.

D4126   Horwich   1962   86A   2/04   P   08896
08896   to Traditional Traction, Wishaw, Warwickshire (despatched from EWS Toton Depot), 8th March 2007; to Severn Valley Railway, November 2009.

D4133   Horwich   1962   36A   7/95   P   08903
08903   to ICI, Billingham, 9th May 1996; to RFS (Engineering) Ltd, Doncaster, for repairs, March 1997; to ICI, Wilton, 4th August 1997; to ICI Billingham, September 1999; to ICI Wilton, 2005; to ICI, Billingham, 17th April 2007.

D4141   Horwich   1962   40B   2/04   P   08911/MATEY
08911   to National Railway Museum, York (despatched from EWS Thornaby Depot), 15th May 2004; to Southall Depot (for use in 'Railway Children' production), 23rd May 2010; to National Railway Museum, York, 8th January 2011; to Southall Depot (for use in 'Railway Children' production), 26th May 2011.

D4142   Horwich   1962   8J   9/02   P   08912
08912   to T.J. Thomson Ltd, Stockton (despatched from EWS Toton Depot), 6th March 2007; to EWS Thornaby Depot, for wheel turning, 1st February 2008; to A.V. Dawson Ltd, Middlesbrough, 20th March 2008; stripped for spares, May 2008.

D4143   Horwich   1962   66B   11/05   P   08913/HYWELL
08913   to LH Group, Barton under Needwood (despatched from EWS Toton Depot), 20th March 2007; to Manchester Ship Canal Company, Barton Dock, on hire, 24th April 2007; purchased by Hunslet Engine Company, March 2008; to Cleveland Potash Ltd, Boulby Mine, on hire, May 2009; to LH Group, Barton under Needwood, 30th October 2009; to Daventry International Rail Freight Terminal, on hire, 10th December 2009; to LH Group, Barton under Needwood, week ending 24th June 2011.

D4145   Horwich   1962   8J   2/04   P   08915
08915   to Traditional Traction, Wishaw, Warwickshire (despatched from EWS Toton Depot), and moved direct to Colne Valley Railway, 13th March 2007; to Stephenson Railway Museum, Chirton, 5th November 2009.

D4154   Horwich   1962   66B   8/06   P   08924
08924   to C.F. Booth Ltd, Rotherham (despatched from DBS, Tyne Yard), 27th

January 2011; purchased by HNRC; to Barrow Hill Engine Shed Society, Staveley, 15th February 2011.

D4156   Horwich   1962   8J   11/96   F   08926
08926   to Traditional Traction, Wishaw, Warwickshire, 28th March 2007; used for spares; remains to European Metal Recycling, Kingsbury, for scrap, 17th June 2007; scrapped, July 2007.

D4157   Horwich   1962   66B   6/05   P   08927
08927   to Traditional Traction, Wishaw, Warwickshire, and moved direct to Gloucestershire Warwickshire Railway, Toddington, 28th March 2007; to Alstom, Wembley, for wheel turning, 19th September 2007; returned to Traditional Traction (GWR), 24th November 2007; to Pontypool & Blaenavon Railway, on hire, 28th April 2010; to Gloucestershire Warwickshire Railway, 31st January 2011; to Southall Depot (for use in Railway Children production), May 2011.

D4158   Darlington   1962   NC   ?   F   08928
08928   sold to Cotswold Rail, Moreton in Marsh, about June 2001; to Mid Norfolk Railway, Dereham (despatched by rail from Crown Point Depot, Norwich), for loading, 14th August 2001; to Brush, Loughborough, for repairs, 15th August 2001; sold to Harry Needle Railroad Company; to Barrow Hill Engine Shed Society, Staveley, 17th July 2003; to HNRC, Long Marston, for storage, 21st March 2006; to European Metal Recycling, Kingsbury, for scrap, 3rd December 2010; scrapped, December 2010.

D4163   Darlington   1962   81A   2/08   P   08933
08933   sold to T.J. Thomson Ltd, Stockton-on-Tees, 2009; resold direct to Foster Yeoman Quarries Ltd, Merehead, Somerset (despatched from Hoo Junction), 12th March 2009; to Knight's Rail Services, Eastleigh Works, 19th October 2010.

D4166   Darlington   1962   31B   12/92   P   08936
08936   to South Yorkshire Railway Preservation Society (HNRC), Meadowhall, Sheffield, 31st January 1994; to Railway Age, Crewe, on hire, 22nd January 1999; to Fragonset, Derby, for overhaul, 9th June 2000; to Barrow Hill Engine Shed Society (HNRC), Staveley, 4th December 2001; to Fragonset, Derby, May 2002; to Barrow Hill Engine Shed Society, 4th February 2003; to Network Rail, Whitemoor Yard, March, on hire, about May 2004; purchased by Cotswold Rail, Moreton in Marsh, about July 2004; to Allelys Ltd, Studley, Warwickshire, for storage, by 31st January 2006; to Alstom, Stonebridge Park, on hire, by 23rd February 2006; to Willesden TMD, 28th May 2006; to Horton Road Depot, Gloucester, for storage, early November 2006; purchased by RMS Locotec, 2007; to RMS Locotec, Wakefield, for repairs, 4th October 2007; to Corus, Shotton, on hire, November 2007.

D4167   Darlington   1962   84A   12/93   P   D4167/
0893   BLUEBELL MEL
to Aggregate Industries Ltd, Meldon Quarry, Devon, 1995; to RMS Locotec, Dewsbury, 25th November 2005; to Wabtec, Doncaster, for sub-contract repairs, May 2006; to Aggregate Industries, Meldon Quarry, 24th April 2007.

D4173   Darlington   1962   1A   7/88   P   08943
08943   to BREL, Crewe Works, April 1989; loco included in sale when works privatised; to ABB Transportation Ltd, York, February 1993; to ABB British Wheelset Ltd, Trafford Park, Manchester, about August 1996; to ABB Transportation Ltd, Crewe, May

1998; purchased by HNRC, July 2009; to Barrow Hill Engine Shed Society, Staveley, 12th February 2010; to Southall Depot (for use in connection with Railway Children play at Waterloo Station), on hire, 25th May 2010; to Bombardier Transportation, Central Rivers Depot, Barton under Needwood, 7th January 2011.

D4174   Darlington   1962   81A   5/98   P   08944
08944   to Mike Darnall, Newton Heath, Manchester, November 2000; to Wabtec, Doncaster, for overhaul, late 2000; to East Lancashire Railway, Bury, by April 2001; to Crewe Electric Depot, for tyre turning, 7th March 2007; to East Lancashire Railway, Bury, 15th March 2007.

D4176   Darlington   1962   8J   2/02   F   08946
08946   to Traditional Traction, Wishaw, Warwickshire, 23rd March 2007; scrapped by Moveright International, June 2008.

D4177   Darlington   1962   81A   3/04   P   08947
08947   sold to T.J. Thomson Ltd, Stockton-on-Tees, 2007; resold direct to Foster Yeoman, Merehead Quarry (despatched from EWS Westbury), 13th April 2007; to Whatley Quarry, by 9th September 2007; to Merehead Quarry, for open day held on 22nd June 2008.

D4184   Darlington   1963   2F   5/04   P   08954
08954   to HNRC; to Boden Rail Engineering, Washwood Heath (despatched from DBS, Toton Depot), for repairs, 15th February 2011.

D4186   Darlington   1963   ?   ?   P   08956
08956   to SERCO Railtest, Derby, about August 2001; to Fragonset Rail, Derby, by May 2005; to SERCO Railtest, Derby, about July 2007; to SERCO, Asfordby, 9th December 2008.

SECTION 17

British Railways built 0-6-0 diesel electric locomotives, numbered D3665-D3671, D3719-D3721, and D4099-D4114. Basically the same as a 08 shunter, but capable of a higher speed (27.5mph instead of 20mph). Later classified TOPS Class 09.

D3665   Darlington   1959   16A   2/09   P   09001
09001   to Peak Rail, Rowsley (despatched from DBS, Doncaster), 28th January 2011.

D3666   Darlington   1959   75C   9/92   P   D3666
09002   to South Devon Railway, Buckfastleigh, 11th June 1993; sold to GBRf, early 2011; to LH Group, Barton under Needwood, for repairs, 2nd March 2011.

D3667   Darlington   1959   81A   5/08   P   09003
09003   to HNRC, purchased July 2010; to Boden Rail Engineering, Washwood Heath (despatched from Margam Depot), for repairs, August 2010.

D3668   Darlington   1959   75C   4/99   P   D3668
09004   to Lavender Line, Isfield, about 7th December 2000; to Spa Valley Railway, Tunbridge Wells, 12th March 2003; to Swindon & Cricklade Railway, 27th June 2009.

D3671   Darlington   1959   16A   12/09   P   09007
09007   to London Overground, Willesden Depot, 23rd September 2010.

D3719   Darlington   1959   87B   5/04   P   09008
09008   to HNRC; to Boden Rail Engineering, Washwood Heath (despatched from Bescot), for repairs, 7th January 2011.

D3720   Darlington   1959   81A   5/04   P   09009
09009   to C.F. Booth Ltd, Rotherham (despatched from DBS, Toton Depot), 25th January 2011; to LH Group, Barton under Needwood, for repairs, 14th February 2011; resold to GBRf.

D3721   Darlington   1959   81A   9/04   P   D3721
09010   to South Devon Railway (despatched from DBS, Hither Green Depot), 30th September 2010.

D4100   Horwich   1961   81A   2/04   P   09012/DICK HARDY
09012   to Barrow Hill Engine Shed Society (HNRC), Staveley (despatched from DBS, Hither Green Depot), September 2010.

D4102   Horwich   1961   16A   3/09   P   09014
09014   to HNRC; to Boden Rail Engineering, Washwood Heath (despatched from DBS, Doncaster), for repairs, 21st January 2011.

D4103   Horwich   1961   87B   3/07   P   09015/ROB
09015   to T.J. Thomson Ltd, Stockton, 21st February 2011; purchased by National Railway Museum, York, but moved direct to Moveright International, Wishaw, for storage and extraction of spares, 29th March 2011.

D4105   Horwich   1961   ?   ?   P   09017/LEO
09017   purchased by National Railway Museum, York; stored at Network Rail, Klondyke Yard, York, 10th August 2011.

D4106   Horwich   1961   81A   7/04   P   09018
09018   to HNRC; to Boden Rail Engineering, Washwood Heath (despatched from DBS, Hither Green Depot), for repairs, 29th September 2010.

D4107   Horwich   1961   16A   6/10   P   09019
09019   to HNRC; to Barrow Hill Engine Shed Society, Staveley (despatched from DBS, Toton Depot), 16th February 2011.

D4113   Horwich   1962   75C   10/05 P   09025
09025   to East Kent Light Railway, Shepherdswell (despatched from Selhurst Depot), about 25th September 2005.

SECTION 18

Clayton Equipment Co Ltd built Bo-Bo diesel electric locomotives, numbered D8500-D8616, and introduced 1962. Fitted with two Paxman 6ZHXL engines (each developing 450bhp at 1500rpm) and driving wheels of 3ft 3½in diameter. Later classified TOPS Class 17, with three sub-divisions of which D8568 was 17/1.

D8568   CE   4365U/69   1963   66A   10/71   P   D8568
to Hemel Hempstead Lightweight Concrete Co Ltd, Cupid Green, Hertfordshire, 11th September 1972; to Ribblesdale Cement Ltd, Clitheroe, Lancashire, from 16th to 24th June 1977; sold to North Yorkshire Moors Railway, December 1982; left Clitheroe on 9th February 1983; arrived at NYMR, Grosmont, 11th February 1983; to BR Gloucester, 2nd August 1991; exhibited at BR Gloucester Depot open day, 18th August 1991; to BR Old Oak Common Depot open day, August 1991; to BR Stonebridge Park Sidings, November 1991; to Willesden Depot, for tyre turning, December 1991; to Chinnor and Princes Risborough Railway, Oxfordshire, 25th April 1992; to Severn Valley Railway, for gala, October 1998; returned to C&PRR; to Old Oak Common Depot, open day, August 2000; returned to C&PRR; to Barrow Hill Engine Shed Society, for open day, October 2001; returned to C&PRR, 9th October 2001.

SECTION 19

British Railways built 0-6-0 diesel hydraulic locomotives, numbered D9500-D9555, and introduced 1964. Fitted with a Paxman 'Ventura' 6YJX engine developing 650bhp at 1500rpm, and driving wheels of 4ft 0in diameter. Later classified TOPS Class 14.

D9500   Swindon   1964   86A   4/69   P   D9500
to NCB Ashington Colliery (despatched from Canton Depot, Cardiff, 17th November 1969), November 1969; to Lambton Engine Works, Philadelphia, 12th July 1978; to Ashington Colliery, 26th June 1979; to BR Thornaby Depot, for attention, by 14th May 1982; returned to Ashington Colliery, by 24th May 1982; to Lambton Engine Works, 12th May 1983; to Ashington Colliery, 25th July 1983; to Llangollen Railway, 25th September 1987; to Heritage Centre, Swindon, April 1988; to West Somerset Railway, 4th November 1989; to South Yorkshire Railway Preservation Society, Meadowhall, Sheffield, August 1992; to Barrow Hill Engine Shed Society, Staveley, 2nd August 2001; to Peak Rail, Rowsley, to be re-engined, 13th May 2008.

D9502   Swindon   1964   86A   4/69   P   D9502
to NCB Ashington Colliery (despatched from Canton Depot, Cardiff, on 30th June 1969), July 1969; to Burradon Colliery, by September 1969; to Ashington Central Workshops, about July 1973; to Burradon Colliery, March 1974; to Backworth Colliery, January 1976; to Weetslade Loco Shed, about June 1976; to Ashington Colliery, 24th April 1981; to Lambton Engine Works, 28th April 1983; to Ashington Colliery, 12th May 1983; to Llangollen Railway, 25th September 1987; to South Yorkshire Railway Preservation Society (HNRC), Meadowhall, Sheffield, 12th March 1992; to Peak Rail, Rowsley, March 2002.

D9503   Swindon   1964   50B   4/68   F   65/ 8411-25
to Stewarts & Lloyds Minerals Ltd, Harlaxton Quarries, Lincolnshire, November 1968; to Corby Quarries, July 1974; scrapped, September 1980.

D9504   Swindon   1964   50B   4/68   P   D9504
to NCB Philadelphia Loco Shed, County Durham, 29th November 1968; to Boldon Colliery, 21st August 1973; to Philadelphia, 7th September 1973; to Boldon Colliery, February 1974; to Backworth Colliery, December 1974; to Burradon Colliery, January 1975; to Weetslade Loco Shed, 5th January 1976; to Lambton Engine Works, Philadelphia, 21st April 1981; to Ashington Colliery, 11th September 1981; to Kent & East Sussex Railway, Tenterden, 26th September 1987; to Nene Valley Railway, Wansford, for overhaul, 25th February 1998; returned to Kent & East Sussex Railway, 21st April 1999; to Channel Tunnel Rail Link, Beechbrook Farm, near Ashford, on hire, by October 2001; to Chatham Dockyard, about February 2003; to Nene Valley Railway, 4th April 2003; to EWS Toton Depot, for wheel turning, 7th January 2004; returned to NVR, 9th January 2004; to Victa Rail, March, on hire, 15th January 2004; to Nene Valley Railway, for repairs, 8th April 2004; to CTRL, Swanscombe, on hire, 15th June 2004; to CTRL, Dagenham, on hire, 21st November 2004; to Nene Valley Railway, for repairs, 18th March 2005; to CTRL, Dagenham, on hire, by 3rd May 2005; to Nene Valley Railway; to CTRL, Swanscombe, on hire, 7th November 2005; to Nene Valley Railway, 13th January 2006; to CTRL, Dagenham, on hire, 2nd March 2006; to Nene Valley Railway, Wansford, 25th January 2007; to Aggregate Industries Ltd, Bardon Hill, on hire, 16th January 2008; to Nene Valley Railway, Wansford, 10th February 2009; to Kent & East Sussex Railway, 5th May 2010.

D9505   Swindon   1964   50B   4/68   F   MICHLOW
to APCM, Hope, Derbyshire, 26th September 1968; sold for export (see Appendix C); left Hope on 5th May 1975.

D9507   Swindon   1964   50B   4/68   F   55/ 8311-35
to Stewarts & Lloyds Minerals Ltd, Corby Quarries, November 1968; to BSC Steelworks Disposal Site, Corby, December 1980; scrapped on site by Shanks & McEwan Ltd of Corby, September 1982.

D9508   Swindon   1964   87E   10/68   F   No.9/ 9312-99
to NCB Ashington Colliery (despatched from BR Canton Depot, Cardiff, on 6th March 1969), March 1969; withdrawn November 1983; scrapped on site by D. Short Ltd of North Shields, 17th January 1984.

D9510   Swindon   1964   50B   4/68   F   60/ 8411-23
to Stewarts & Lloyds Minerals Ltd, Buckminster Quarries, Lincolnshire, December 1968; to Corby Quarries, June 1972; to BSC Tube Works, Corby, January 1981; scrapped on site by Shanks & McEwan Ltd of Corby, August 1982.

D9511   Swindon   1964   50B   4/68   F   9312-98
to NCB Ashington Colliery, January 1969; to Bates Colliery, Blyth, April 1969; to Burradon Colliery, by 25th May 1969; to Ashington Colliery, about October 1972; dismantled for spares after a fire; remains scrapped, July 1979.

D9512   Swindon   1964   50B   4/68   F   63/ 8411-24
to Stewarts & Lloyds Minerals Ltd, Buckminster Quarries, Lincolnshire, December 1968; to Corby Quarries, 6th September 1972; used for spares; to BSC Steelworks Disposal Site, Corby, 29th December 1980; scrapped, about February 1982.

D9513   Swindon   1964   86A   3/68   P   NCB 38
to Arnott Young Ltd, Parkgate, Rotherham (despatched from BR Worcester Depot), 18th

July 1968; to Hargreaves Industrial Services Ltd, NCBOE British Oak Disposal Point, Crigglestone, November 1968; to NCBOE Bowers Row Disposal Point, Astley, 5th September 1969; sold to NCB North East Area; to Allerton Bywater Central Workshops, West Yorkshire, for overhaul, October 1973; to Ashington Colliery, January 1974; to Backworth Colliery, July 1974; to Burradon Colliery, July 1974; to Backworth Colliery, 5th January 1976; to Lambton Engine Works, Philadelphia, 22nd November 1976; to Ashington Colliery, 14th February 1977; to Lambton Engine Works, 8th September 1977; to Ashington Colliery, 14th November 1977; to Lambton Engine Works, 30th June 1982; to Ashington Colliery, 17th February 1983; sold to C.F. Booth Ltd of Rotherham, summer 1987; did not leave site and was re-sold by Booth's for preservation; to Embsay & Bolton Abbey Railway, 12th October 1987.

D9514   Swindon   1964   86A   4/69   F   No.4/ 9312-96
to NCB Ashington Colliery (despatched from Canton Depot, Cardiff, on 30th June 1969), July 1969; to BR Gosforth Depot, for repairs, 11th October 1975; to Ashington Colliery, November 1975; to Lambton Engine Works, Philadelphia, about August 1977; to Ashington Colliery, about 1978; to Lambton Engine Works, September 1980; returned to Ashington Colliery; to BR Thornaby Depot, for repairs, October 1982; returned to Ashington Colliery; scrapped on site, November 1985.

D9515   Swindon   1964   50B   4/68   F   62/ 8411-22
to Stewarts & Lloyds Minerals Ltd, Buckminster Quarries, Lincolnshire, 2nd November 1968; to Corby Quarries, September 1972; to BSC Steelworks Disposal Site, Corby, 29th December 1980; to Hunslet Engine Co Ltd, Leeds, December 1981; overhauled and converted to 5ft 6in gauge; exported to Spain from Goole Docks, June 1982; scrapped, by 2002.

D9516   Swindon   1964   50B   4/68   P   D9516
to Stewarts & Lloyds Minerals Ltd, Corby Quarries, November 1968; to BSC Steelworks Disposal Site, Corby, 29th December 1980; to Great Central Railway, Loughborough, 17th October 1981; to Severn Valley Railway diesel weekend, 15th October 1988; returned to GCR; to Nene Valley Railway, Wansford, 8th December 1988; to Boden Rail Engineering, Washwood Heath, 1st April 2011; to Wensleydale Railway, Leeming Bar, 11th April 2011.

D9517   Swindon   1964   86A   10/68   F   No.8/ 9312-93
to NCB Ashington Colliery (despatched from BR Canton Depot, Cardiff, on 17th November 1969), November 1969; to Lambton Engine Works, Philadelphia, 14th June 1977; to Ashington Colliery, 5th September 1977; withdrawn November 1983; scrapped on site by D. Short Ltd of North Shields, January 1984.

D9518   Swindon   1964   86A   4/69   P   No.7/9312-95
to NCB Ashington Colliery (despatched from Canton Depot, Cardiff, on 30th June 1969), July 1969; to Lambton Engine Works, Philadelphia, June 1975; to Ashington Colliery, September 1975; to Lambton Engine Works, 5th September 1980; to Ashington Colliery, 3rd December 1980; to Rutland Railway Museum, Cottesmore, 26th September 1987; to Nene Valley Railway, Wansford, 8th September 2006.

D9520   Swindon   1964   50B   4/68   P   D9520/BSC 45
to Stewarts & Lloyds Minerals Ltd, Glendon Quarries, Northamptonshire, 16th December 1968; to Corby Quarries, 12th January 1970; to BSC Tube Works, Corby, October 1980; to North Yorkshire Moors Railway, 16th March 1981; to Rutland Railway Museum,

Cottesmore, 21st February 1984; to Great Central Railway, Loughborough, on loan, 5th October 1985; returned to Cottesmore, 2nd December 1985; to Great Central Railway, Ruddington, Nottingham, 6th March 1998; to Nene Valley Railway, Wansford, 21st April 2004; to West Somerset Railway, for gala, 15th June 2007; returned to Nene Valley Railway, Wansford; to Appleby Frodingham RPS, Scunthorpe, for gala, 9th to 11th May 2008; to National Railway Museum, York, May 2008; to Barrow Hill Engine Shed Society, Staveley, for gala, August 2008; to Lafarge, Hope, for open day, 6th September 2008; to Nene Valley Railway, 2008; to West Somerset Railway, for gala, June 2009; to Nene Valley Railway, 14th June 2009; to West Somerset Railway, for gala, 7th June 2010; to Nene Valley Railway, 14th June 2010.

D9521   Swindon   1964   87E   4/69   P   D9521
to NCB Ashington Colliery, March 1970; to BR Gosforth Depot for repairs, 11th March 1976; to Ashington Colliery, 17th April 1976; to Lambton Engine Works, Philadelphia, 30th November 1977; to Ashington Colliery, by October 1978; to Lambton Engine Works, 7th January 1982; to Ashington Colliery, 30th June 1982; to Lambton Engine Works, March 1983; to Ashington Colliery, March 1983; sold to C.F. Booth Ltd of Rotherham, summer 1987; did not leave site and was re-sold by Booth's for preservation; to Rutland Railway Museum, Cottesmore, 14th October 1987; to Swanage Railway Society, Dorset, 29th January 1992; to BR Wimbledon Depot, for wheel turning, 8th June 1995; returned to Swanage Railway, 12th June 1995; to Barry Island Railway, Barry Island, 11th November 2004; to Mid Norfolk Railway, 20th May 2008; to Quainton Railway Society, Buckinghamshire, 9th July 2008; to Barry Island Railway, autumn 2008; to Dean Forest Railway, 16th January 2009; to Swindon & Cricklade Railway, for gala, September 2009; returned to Dean Forest Railway, October 2009; to Gwili Railway, on loan, 10th August 2010; to Llangollen Railway, on loan, 25th August 2010; returned to Dean Forest Railway, by 24th April 2011.

D9523   Swindon   1964   50B   4/68   P   D9523
to Stewarts & Lloyds Minerals Ltd, Glendon East Quarries, Northamptonshire, 16th December 1968; to Corby Quarries, 28th May 1980; to BSC Steelworks Disposal Site, 29th December 1980; to Great Central Railway, Loughborough, 17th October 1981; to Nene Valley Railway, Wansford, 7th December 1988; to Boden Rail Engineering, Washwood Heath, 8th April 2011; to Derwent Valley Railway, 21st April 2011.

D9524   Swindon   1964   87E   4/69   P   14901
to BP Refinery Ltd, Grangemouth (despatched from Canton Depot, Cardiff), July 1970; fitted with a Dorman type 8QT 500hp engine, by Andrew Barclay, Sons & Co Ltd; to BR Grangemouth Depot, for repairs, November 1971; returned to BP; to BR Eastfield Depot, for repairs, March 1978; returned to BP; to Scottish Railway Preservation Society, Falkirk, 9th September 1981; to SRPS, Bo'ness, 7th February 1988; fitted with Rolls-Royce type DV8 750hp engine; sold to Middle Peak Railways, about June 2006; to RMS Locotec, Wakefield, for overhaul, 1st July 2006; to Elsecar Steam Railway, near Barnsley, 17th April 2007; to Midland Railway Centre, for gala, 18th May 2010; to Peak Rail, Rowsley, 1st July 2010; to Gwili Railway, on hire, 2nd April 2011.

D9525   Swindon   1964   50B   4/68   P   D9525
to NCB Philadelphia Loco Shed, County Durham, 28th November 1968; to Burradon Colliery, 7th March 1975; to Ashington Colliery, 14th March 1975; to Backworth Colliery, 13th December 1975; to Ashington Colliery, August 1980; to Weetslade Loco Shed, January 1981; to Ashington Colliery, 24th April 1981; to Lambton Engine Works,

Philadelphia, 25th July 1983; to Ashington Colliery, 19th December 1983; to Kent & East Sussex Railway, Tenterden, 29th September 1987; to Great Central Railway, Ruddington, Nottingham, 26th June 2000; to Barrow Hill Engine Shed Society, Staveley, 13th August 2001; to Battlefield Line, Shackerstone, 2nd March 2002; to South Devon Railway, 27th May 2004; to Peak Rail, Rowsley, 13th April 2005.

D9526   Swindon   1964   86A   11/68   P   D9526
to APCM, Westbury, Wiltshire, January 1970; to APCM, Dunstable, 28th May 1971; to APCM, Westbury, 24th November 1971; to West Somerset Railway, Minehead, 3rd April 1980.

D9527   Swindon   1965   86A   4/69   F   No.6/ 9312-94
to NCB Ashington Colliery (despatched from Canton Depot, Cardiff, on 30th June 1969), July 1969; to Lambton Engine Works, Philadelphia, 28th May 1977; to Ashington Colliery, 20th September 1977; to Lambton Engine Works, 5th September 1978; to Ashington Colliery, October 1978; withdrawn November 1983; scrapped on site by D. Short Ltd of North Shields, January 1984.

D9528   Swindon   1965   86A   3/69   F   No.2/ 9312-100
to NCB Ashington Colliery (despatched from Canton Depot, Cardiff, on 6th March 1969), March 1969; to BR Gosforth Depot, for repairs, 11th March 1976; to Ashington Colliery by May 1976; to Lambton Engine Works, Philadelphia, June 1977; to Ashington Colliery, by November 1977; scrapped, December 1981.

D9529   Swindon   1965   50B   4/68   P   14029
to Stewarts & Lloyds Minerals Ltd, Buckminster Quarries, Lincolnshire, August 1968; to Corby Quarries, September 1972; to BSC Steelworks Disposal Site, Corby, 29th December 1980; to North Yorkshire Moors Railway, 16th March 1981; to Great Central Railway, Loughborough, 11th December 1984; to BR Coalville Depot, open day, 5th June 1988; returned to GCR; to Nene Valley Railway, Wansford, 8th December 1988; to Battlefield Line, Shackerstone, on loan, 10th April 1995; to Nene Valley Railway, 9th November 1995; to Kent & East Sussex Railway, Tenterden, Kent, 23rd June 2000; to Channel Tunnel Rail Link, Beechbrook Farm, near Ashford, on hire, week commencing 16th July 2001; to Nene Valley Railway, for repairs, April 2002; returned to CTRL, on hire, 28th June 2002; to Tilbury Docks, 2nd January 2003; to ARC Stone, Chequers Lane, Dagenham, on hire, 19th July 2004; to Nene Valley Railway, 19th September 2004; to EWS Toton Depot, for tyre turning, 17th November 2004; to NVR, 29th November 2004; to CTRL, Dagenham, on hire, 23rd February 2005; to CTRL, Swanscombe, on hire, September 2005; returned to Nene Valley Railway, 14th April 2006; to CTRL, Dagenham, on hire, 7th September 2006; to Nene Valley Railway, for repairs, 29th September 2006; to CTRL, Swanscombe, on hire, 24th January 2007; to Nene Valley Railway, for repairs, 24th April 2007; to Kent & East Sussex Railway, Tenterden, 30th May 2007; to Nene Valley Railway, 29th December 2008; to Aggregate Industries Ltd, Bardon Hill, on hire, 7th January 2009; to Nene Valley Railway, 9th October 2010.

D9530   Swindon   1965   86A   10/68   F   NFT
to Gulf Oil Co Ltd, Waterston, Pembrokeshire, 26th September 1969; to BR Swindon Works, for overhaul, 5th August 1971; returned to Gulf Oil, 7th October 1971; to NCB Mardy Colliery, Glamorgan, via BR Canton Depot, Cardiff, after 16th November 1975; to BR Canton Depot and BR Ebbw Junction Depot, for repairs, July to 16th August 1976;

to Mardy Colliery, August 1976; to BR Canton Depot, Cardiff, 2nd October 1977; to Mardy Colliery, about December 1977; scrapped, March 1982.

D9531   Swindon   1965   86A   12/67   P   D9531
to Arnott Young Ltd, Parkgate, Rotherham (despatched from BR Worcester), 18th July 1968; to Hargreaves Industrial Services Ltd, NCBOE British Oak Disposal Point, Crigglestone, November 1968; to NCB Burradon Colliery, 10th October 1973; to Ashington Colliery, by March 1974; to Lambton Engine Works, Philadelphia, 26th September 1981; to Ashington Colliery, 27th October 1981; sold to C.F. Booth Ltd of Rotherham, summer 1987; did not leave site and was re-sold by Booth's for preservation; to East Lancashire Railway, Bury, 3rd October 1987.

D9532   Swindon   1965   50B   4/68   F   57/ 8311-37
to Stewarts & Lloyds Minerals Ltd, Corby Quarries, November 1968; to BSC Steelworks Disposal Site, Corby, 29th December 1980; scrapped on site by Shanks & McEwan Ltd of Corby, February 1982.

D9533   Swindon   1965   50B   4/68   F   47/ 8311-26
to Stewarts & Lloyds Minerals Ltd, Corby Quarries, December 1968; to BSC Steelworks Disposal Site, Corby, 29th December 1980; scrapped on site by Shanks & McEwan Ltd of Corby, September 1982.

D9534   Swindon   1965   50B   4/68   F   ECCLES
to APCM, Hope, Derbyshire, October 1968; sold for export (see Appendix C); left Hope on 5th May 1975.

D9535   Swindon   1965   86A   12/68   F   37/ 9312-59
to NCB Ashington Colliery, November 1970; to Burradon Colliery, January 1971; to Ashington Central Workshops, about July 1973; to Burradon Colliery, about March 1974; to Weetslade Loco Shed, 5th January 1976; to Backworth Colliery, about May 1976; to Ashington Colliery, 13th September 1980; to Lambton Engine Works, Philadelphia, 5th December 1980; to Ashington Colliery, 23rd April 1981; withdrawn November 1983; scrapped on site by D. Short Ltd of North Shields, January 1984.

D9536   Swindon   1965   87E   4/69   F   No.5/ 9312-91
to NCB Ashington Colliery (despatched from Canton Depot, Cardiff), March 1970; to BR Gosforth Depot, for wheel turning, August 1973; returned to Ashington Colliery; to Lambton Engine Works, Philadelphia, 28th January 1977; to Ashington Colliery, 15th May 1977; to Lambton Engine Works, January 1978; to Ashington Colliery, June 1978; to Lambton Engine Works, 16th September 1981; to Ashington Colliery, January 1982; scrapped on site, week ending 30th November 1985.

D9537   Swindon   1965   50B   4/68   P   D9537
to Stewarts & Lloyds Minerals Ltd, Corby Quarries, November 1968; to BSC Penn Green Crane Depot, for storage, about November 1981; to Gloucestershire Warwickshire Railway Society, Toddington, 23rd November 1982; to Kent Loco Group, Rippingale Station, Lincolnshire, for storage, 8th May 2003.

D9538   Swindon   1965   87E   4/69   F   160
to Shell-Mex & BP Ltd, Shell Haven, Essex, April 1970; returned to BR Swindon Works, for overhaul, by 5th August 1970; resold to British Steel Corporation; to BSC Ebbw Vale Steelworks, Monmouthshire, 22nd February 1971; to BSC Corby Quarries, April 1976;

scrapped on site by Shanks & McEwan Ltd of Corby, September 1982.

D9539   Swindon   1965   50B   4/68   P   D9539
to Stewarts & Lloyds Minerals Ltd, Corby Quarries, October 1968; to BSC Steelworks Disposal Site, Corby, 29th December 1980; to Gloucestershire Warwickshire Railway, Toddington, 23rd February 1983; to Ribble Steam Railway, Preston, 26th July 2005; to EWS Crewe Depot, for tyre turning, 16th February 2009; returned to Ribble, 27th April 2009.

D9540   Swindon   1965   50B   4/68   F   36/ 508/ 2233-508
to NCB Philadelphia Loco Shed, County Durham, 29th November 1968; to Burradon Colliery, 25th November 1971; to Ashington Colliery, June 1972; to Burradon Colliery, by April 1974; to Weetslade Loco Shed, 5th January 1976; to Ashington Colliery, 24th April 1981; withdrawn November 1983; scrapped on site by D. Short Ltd of North Shields, 11th January 1984.

D9541   Swindon   1965   50B   4/68   F   66
to Stewarts & Lloyds Minerals Ltd, Harlaxton Quarries, Lincolnshire, November 1968; to Corby Quarries, August 1974; to BSC Steelworks Disposal Site, Corby, 29th December 1980; scrapped on site by Shanks & McEwan Ltd of Corby, August 1982.

D9542   Swindon   1965   50B   4/68   F   48/ 8311-27
to Stewarts & Lloyds Minerals Ltd, Corby Quarries, December 1968; to BSC Steelworks Disposal Site, Corby, 29th December 1980; scrapped on site by Shanks & McEwan Ltd of Corby, August 1982.

D9544   Swindon   1965   50B   4/68   F   D9544
to Stewarts & Lloyds Minerals Ltd, Corby Quarries, 2nd November 1968; dismantled and used for spares in 1970; remains scrapped, September 1980.

D9545   Swindon   1965   50B   4/68   F   D9545
to NCB Ashington Colliery, November 1968; later dismantled and used for spares; remains scrapped, early July 1979.

D9547   Swindon   1965   50B   4/68   F   49/ 8311-28
to Stewarts & Lloyds Minerals Ltd, Corby Quarries, December 1968; to BSC Steelworks Disposal Site, Corby, 29th December 1980; scrapped on site by Shanks & McEwan Ltd of Corby, August 1982.

D9548   Swindon   1965   50B   4/68   F   67/ 8411-27
to Stewarts & Lloyds Minerals Ltd, Harlaxton Quarries, Lincolnshire, November 1968; to Corby Quarries, August 1974; to BSC Steelworks Disposal Site, Corby, 29th December 1980; to Hunslet Engine Co Ltd, Leeds, 19th November 1981; overhauled and rebuilt to 5ft 6in gauge; exported to Spain via Goole Docks, June 1982.

D9549   Swindon   1965   50B   4/68   F   64/ 8311-33
to Stewarts & Lloyds Minerals Ltd, Corby Quarries, November 1968; to Glendon East Quarries, October 1973; to Corby Quarries, 26th June 1974; to BSC Tube Works, Corby, September 1980; to Hunslet Engine Co Ltd, Leeds, 14th November 1981; overhauled and rebuilt to 5ft 6in gauge; exported to Spain via Goole Docks, June 1982; scrapped at Tarragona, Spain, March 2007.

D9551   Swindon   1965   50B   4/68   P   D9551
to Stewarts & Lloyds Minerals Ltd, Corby Quarries, December 1968; to BSC Tube Works, Corby, July 1980; to West Somerset Railway, Minehead, 5th June 1981; to Royal Deeside Railway, Banchory, near Aberdeen, 6th to 9th November 2000.

D9552   Swindon   1965   50B   4/68   F   59/ 8411-21
to Stewarts & Lloyds Minerals Ltd, Buckminster Quarries, Lincolnshire, September 1968; to Corby Quarries, June 1972; scrapped, September 1980.

D9553   Swindon   1965   50B   4/68   P   D9553/54
to Stewarts & Lloyds Minerals Ltd, Corby Quarries, November 1968; to BSC Steelworks Disposal Site, Corby, 29th December 1980; to Gloucestershire Warwickshire Railway, Toddington, 23rd February 1983.

D9554   Swindon   1965   50B   4/68   F   58/ 8311-38
to Stewarts & Lloyds Minerals Ltd, Corby Quarries, November 1968; to BSC Steelworks Disposal Site, Corby, 29th December 1980; scrapped on site by Shanks & McEwan Ltd of Corby, August 1982.

D9555   Swindon   1965   87E   4/69   P   D9555
to NCB Burradon Colliery (despatched from Canton Depot, Cardiff), March 1970; to Ashington Colliery, 7th February 1975; to Burradon Colliery, March 1975; to Backworth Colliery, 3rd December 1975; to Ashington Colliery, 15th August 1980; to Rutland Railway Museum, Cottesmore, 24th September 1987; to Northampton & Lamport Railway, on loan, August 1998; to Rutland Railway Museum, Cottesmore, 27th October 1998; to Old Oak Common Depot, open day, August 2000; returned to Cottesmore; to Dean Forest Railway, on loan, about March 2002; to EWS Toton Depot, for tyre turning, 10th June 2003; returned to Dean Forest Railway, 11th June 2003.

SECTION 20

LMSR and British Railways built 0-6-0 diesel electric locomotives, numbered 12033-12138, and introduced 1945. Fitted with an English Electric 6KT engine developing 350bhp at 680rpm, and driving wheels of 4ft 0½in diameter. Later classified TOPS Class 11.

12049   Derby   1948   1E   10/71   F   12049
to Day & Sons (Brentford) Ltd, Brentford Town Goods Depot, London, October 1972; to BR Old Oak Common Depot, for repairs, October 1976; returned to Day & Sons, January 1977; to Mid-Hants Railway, Ropley, Hampshire, 28th July 1998; suffered from fire damage, 26th July 2010; to European Metal Recycling, Kingsbury, for scrap, 3rd November 2010; scrapped July 2011.

12050   Derby   1949   9A   7/70   F   12050
to NCB Philadelphia Loco Shed, County Durham (despatched from Newton Heath), April 1971; dismantled for spares, June 1971; remains scrapped, June 1972.

12052   Derby   1949   5A   6/71   P   MP228
to Derek Crouch (Contractors) Ltd, NCBOE Widdrington Disposal Point, Northumberland, 14th December 1971; to Scottish Industrial Railway Centre, Dunaskin, Dalmellington, 2nd October 1988; to SIRC, Minnivey, 8th May 1994; to SIRC Dunaskin, 11th March 2002; to Caledonian Railway, Brechin, about April 2002.

12054   Derby   1949   6A   7/70   F   12054
to A.R. Adams & Son, Newport, September 1971; used as a hire loco (see Appendix A); scrapped, April 1984.

12060   Derby   1949   9A   2/71   F   512/ 2233-512
to NCB Derwenthaugh Loco Shed, Blaydon, County Durham, March 1971; to Philadelphia Loco Shed, April 1971; scrapped on site by C.F. Booth Ltd of Rotherham, November 1985.

12061   Derby   1949   8J   10/71 P   NPT
to NCB Nantgarw Coking Plant, Glamorgan (despatched from BR Springs Branch, Wigan), 11th December 1972; to BR Canton Depot, Cardiff, for repairs, 5th December 1974; to Nantgarw, December 1974; to BR Swindon Works, for repairs, 10th December 1981; to Nantgarw, 8th February 1982; to Vale of Neath Railway Society, Aberdulais, 23rd August 1987; to Gwili Railway Company, Bronwydd Arms, 13th September 1991; to Peak Rail, Rowsley, 11th June 2004.

12063   Derby   1949   8F   1/72   F   5
to NCB Nantgarw Coking Plant, Glamorgan, 11th December 1972; to BR Canton Depot, Cardiff, for repairs, January 1977; returned to Nantgarw; scrapped, November 1987.

12071   Derby   1950   8F   10/71 F   6
to NCB Nantgarw Coking Plant, Glamorgan, 11th December 1972; to BR Canton Depot, Cardiff, for repairs, by 23rd December 1974; to Nantgarw, December 1974; to Canton Depot, Cardiff, for repairs, about November 1976; to Nantgarw, 15th November 1976; to BR Ebbw Junction Depot, for repairs, by 21st March 1977; to BR Swindon Works, for repairs, 6th July 1977; to Nantgarw, 2nd October 1977; to BR Swindon Works, for repairs, September 1980; to Nantgarw, February 1981; to NSF Aberaman Phurnacite Plant, July 1987; to C.F. Booth Ltd, Rotherham, 18th July 1990; to South Yorkshire Railway Preservation Society, Meadowhall, Sheffield, 24th August 1992; dismantled for spares, June 1995; remains scrapped at Coopers (Metals) Ltd, Sheffield, June 1995.

12074   Derby   1950   6A   1/72   F   12074
to Johnsons (Chopwell) Ltd, NCBOE, Swalwell Disposal Point, County Durham (despatched from BR Crewe), June 1972; to South Yorkshire Railway Preservation Society (HNRC), Meadowhall, Sheffield, 21st June 1989; to European Metal Recycling, Kingsbury, about June 2001; scrapped, July 2002.

12077   Derby   1950   8F   10/71 P   12077
to Cashmore Ltd, Great Bridge, Staffordshire, September 1973; to Midland Railway, Butterley, Derbyshire, 16th December 1978.

12082   Derby   1950   6G   10/71 P   12049
to Shellstar (UK) Ltd, Ince Marshes, Ellesmere Port (despatched from BR Chester), 27th March 1973; to Manchester Ship Canal Company, Ellesmere Port, on loan, 16th July 1974; returned to Shellstar, 25th October 1974; to BR Swindon Works, for repairs, February 1978; to Shellstar, March 1978; to South Yorkshire Railway Preservation Society (HNRC), Meadowhall, Sheffield, December 1991; to Cobra Railfreight, Wakefield, West Yorkshire, on hire, March 1993; returned to SYRPS, 9th December 1994; to RFS (Engineering) Ltd, Doncaster, for repairs, 28th November 1995; to Cobra Railfreight, Wakefield, on hire, 5th April 1996; to RFS, Doncaster, for repairs, 15th October 1997; to SYRPS, 18th December 1997; to Barrow Hill Engine Shed Society

(HNRC), Staveley, 24th June 1999; Railtrack registered 01553; to Wabtec, Doncaster, 21st December 2000; to Lafarge Cement, Hope, Derbyshire, on hire, 6th August 2003; to BHESS, by May 2004; to Whitemoor Yard, on hire, 29th September 2004; but unsuitable and returned to BHESS, 30th September 2004; to Midland Railway, Butterley, 2004; to HNRC, Long Marston, July 2005; to Deanside Transit, Glasgow, on hire, January 2008; to Barrow Hill Engine Shed Society (HNRC), September 2010; to Mid - Hants Railway, Ropley, 1st November 2010; re-numbered 12049 to replace their original 12049 which suffered fire damage.

12083   Derby   1950   12A   10/71   P   12083/ M413
to Tilcon Ltd, Swinden Lime Works, Grassington, July 1973; to BR Doncaster Depot, for repairs, September 1974; returned to Tilcon Ltd, October 1974; to South Yorkshire Railway Preservation Society (HNRC), Meadowhall, Sheffield, 21st May 1998; to Battlefield Line, Shackerstone, 1st August 2001.

12084   Derby   1950   5A   5/71   F   514/ 2233-514
to NCB Burradon Colliery, Northumberland, October 1971; to Philadelphia Loco Shed, County Durham, 25th November 1971; to Silksworth Colliery, April 1972; to Hylton Colliery, June 1972; to Philadelphia Loco Shed, March 1975; to Easington Colliery, December 1975; to Blackhall Colliery, January 1976; to Bates Colliery, Blyth, April 1976; to Lambton Engine Works, Philadelphia, 25th February 1982; to Philadelphia Loco Shed, 21st October 1983; scrapped on site by C.F. Booth Ltd of Rotherham, November 1985.

12085   Derby   1950   12A   5/71   F   12085
to Thos. W. Ward Ltd, Barrow-in-Furness, May 1973; scrapped, about June 1976.

12088   Derby   1951   8J   5/71   P   12088
to Johnsons (Chopwell) Ltd, NCBOE Swalwell Disposal Point, County Durham (despatched from BR Springs Branch, Wigan), July 1972; to South Yorkshire Railway Preservation Society (HNRC), Meadowhall, Sheffield, June 1989; to Johnson Ltd, Widdrington, on hire, 28th May 1996.

12093   Derby   1951   5A   5/71   P   12093/ MP229
to Derek Crouch (Contractors) Ltd, NCBOE Widdrington Disposal Point, Northumberland, 15th December 1971; to Scottish Industrial Railway Centre, Dunaskin, Dalmellington, 9th October 1988; to Caledonian Railway, Brechin, about April 2002.

12098   Derby   1952   8F   2/71   F   12098
to NCB Derwenthaugh Loco Shed, County Durham, March 1971; to Philadelphia Loco Shed, April 1971; to National Smokeless Fuels Ltd, Lambton Coking Plant, about July 1985; to Stephenson Railway Museum, Middle Engine Lane, North Shields, 5th January 1987; to South Yorkshire Railway Preservation Society (HNRC), Meadowhall, Sheffield, 9th December 1997; to European Metal Recycling, Kingsbury, about June 2001; scrapped, June 2006.

12099   Derby   1952   1E   7/71   P   12099
to Murphy Bros Ltd, NCBOE Lion Disposal Point, Blaenavon, April 1972; to Taylor Woodrow Construction Ltd, NCBOE Cwm Bargoed Disposal Point, 23rd October 1975; to Hargreaves Industrial Services Ltd, NCBOE British Oak Disposal Point, Crigglestone, August 1981; to NCBOE Bowers Row Disposal Point, 11th February 1983; to C.F. Booth Ltd, Rotherham, 26th February 1989; to Severn Valley Railway, 26th March 1990.

12119   Darlington   1952   50B   11/68   F   509/ 2233-509
to NCB Philadelphia Loco Shed, County Durham, 7th February 1969; to Lambton Engine Works, Philadelphia, November 1980; to Philadelphia Loco Shed, January 1981; scrapped on site by C.F. Booth Ltd of Rotherham, November 1985.

12120   Darlington   1952   50B   12/68   F   510
to NCB Philadelphia Loco Shed, County Durham, 7th February 1969; to Whittle Colliery, June 1978; to Lambton Engine Works, Philadelphia, August 1979; scrapped, March 1980.

12122   Darlington   1952   40B   7/71   F   12122
to Murphy Bros Ltd, NCBOE Lion Disposal Point, Blaenavon, 30th January 1972; suffered collision damage, February 1972; to Taylor Woodrow Construction Ltd, NCBOE Cwm Bargoed Disposal Point, October 1975; to Hargreaves Industrial Services Ltd, NCBOE British Oak Disposal Point, Crigglestone (for spares only), August 1981; scrapped on site by Rawden's of Barnsley, October 1985.

12131   Darlington   1952   30A   3/69   P   12131
to NCB Betteshanger Colliery, Kent, 25th March 1969; to Snowdown Colliery, Kent, 22nd June 1976; to North Norfolk Railway, Sheringham, 25th April 1982.

12133   Darlington   1952   40B   1/69   F   511/ 2100-526
to NCB Philadelphia Loco Shed, County Durham, 9th May 1969; to Lambton Engine Works, Philadelphia, 1979; to Whittle Colliery, about March 1981; to Lambton Engine Works, 23rd April 1981; to Philadelphia Loco Shed, 13th August 1981; scrapped on site by C.F. Booth Ltd of Rotherham, November 1985.

SECTION 21

British Railways built 0-6-0 diesel electric locomotives, numbered 15211-15236, and introduced 1949. Fitted with an English Electric 6KT engine developing 350bhp at 680rpm, and driving wheels of 4ft 6in diameter. Later classified TOPS Class 12.

15222   Ashford   1949   73C   10/71   F   15222
to Cashmore Ltd, Newport, June 1972; to John Williams Ltd, Blaenyfan, Kidwelly, 1974, where used as a stationary generator; extant in March 1978, but scrapped later that year.

15224   Ashford   1949   75C   10/71   P   15224
to NCB Betteshanger Colliery, Kent (despatched from BR Brighton), by 15th October 1972; to Snowdown Colliery, Kent, 27th May 1976; left Snowdown Colliery, 9th October 1982; stored in BR Hove Goods Yard; to Brighton Works Locomotive Association, Preston Park Car Sheds, Brighton, April 1983; to Lavender Line, Isfield, East Sussex, June 1985; to Spa Valley Railway, Tunbridge Wells, 21st January 1998 to 1st February 1998.

15231   Ashford   1951   73F   10/71   F   TILCON
to Tilcon Ltd, Swinden Lime Works, Grassington, June 1972; scrapped, January 1984.

SECTION 22

Ruston & Hornsby Ltd built, 3ft 0in gauge, 4-wheel diesel mechanical locomotive, number ED10, built in 1958 (Ruston's class 48DS). Fitted with a Ruston 4YC engine developing 48bhp at 1375rpm, three speed gearbox, and driving wheels of 2ft 6in diameter. No TOPS classification.

ED10   RH   411322   1958   BSD   2/65   P   E9
to Thos. W. Ward Ltd, Sheffield, February 1965; to Cleveland Bridge & Engineering Co Ltd, Darlington, May 1966; used on the Tinsley Viaduct, Sheffield, contract; to Shephard Hill & Co Ltd (contractors), February 1970; fitted with rubber tyres and stabilisers and used on a contract to construct three miles of hover-train track for Tracked Hovercraft Ltd, Earith, Huntingdonshire, from 1971; site closed 7th September 1974; to E. Hampton, Church Farm, Fenstanton, St Ives, Huntingdonshire, for preservation, 1975; to Irchester Narrow Gauge Railway Trust, Irchester Goods Shed, Northamptonshire, 28th September 1987; to Irchester Country Park, Northamptonshire, 8th June 1988; re-gauged to metre gauge, 5th May 1991.

SECTION 23

Ruston & Hornsby Ltd built, 1ft 6in gauge, 4-wheel diesel mechanical locomotive, number ZM32, built 1957 (Ruston's class LAT). Fitted with a Ruston 2VSH engine developing 20bhp at 1200rpm, two speed gearbox, and driving wheels of 1ft 4¼in diameter. No TOPS classification.

ZM32   RH   416214   1957   ZJ   3/64   P   ZM32/ HORWICH/11
to S.E.E.C. Manchester, September 1965; sold to a buyer in British Honduras, but sale cancelled and loco stored at Liverpool Docks; resold for preservation; to R.P. Morris, Longfield, Kent, December 1971; to Alan Keef Ltd, Cote, Oxfordshire, 17th April 1973; rebuilt to 2ft 0in gauge; to Narrow Gauge Railway Centre of North Wales, Gloddfa Ganol, Blaenau Ffestiniog, Gwynedd, 20th July 1976; to R.P. Morris, Blaenau Ffestiniog, 1998; to FMB Engineering, Oakhanger, Hampshire (for repairs and conversion back to 1ft 6in gauge), 6th October 1999; to Uppertown, near Ashover, for storage, 30th May 2000; to Steeple Grange Light Railway, Wirksworth, Derbyshire, 9th June 2000; to Whaley Bridge, for overhaul, 24th November 2002; to Steeple Grange Light Railway, Wirksworth, Derbyshire, 15th January 2003.

SECTION 24

English Electric Ltd built, 0-6-0 diesel electric locomotives, built 1956. Fitted with an English Electric 6RKT engine developing 500bhp at 750rpm, and driving wheels of 4ft 0in diameter. These locomotives were tested by British Railways but were never incorporated into capital stock. No TOPS classification.

D0226   EE   2345   1956   - -   12/60   P   D226/ VULCAN
**   VF   D226**

Given trials on British Railways from 1956 to December 1960; returned to English Electric Ltd, Vulcan Works, Newton-le-Willows, January 1961; stored; to Keighley & Worth Valley Railway, Haworth, March 1966; to BR Doncaster Depot, for wheel turning, 4th March 1979; returned to KWVR, Haworth, 1979; to Railfest, York, for display, 26th May 2004; returned to KWVR, June 2004.

D0227   EE   2346   1956   - -   9/59   F   D0227/ BLACK PIG
**   VF   D227**

Given trials on British Railways from 1956 to 1960; returned to English Electric Ltd, 1960; to Robert Stephenson & Hawthorns Ltd, Darlington 1960; scrapped, July 1964.

SECTION 25

Ruston & Hornsby Ltd built, 4-wheel diesel mechanical locomotives (Ruston's class LB). Fitted with Ruston 3VSH engines developing 31bhp at 1800rpm, and 1ft 4¼in diameter wheels. Ruston & Hornsby built no less than 557 examples of this class, to 23 different narrow gauges, with the two examples recorded below being of 2ft 0in gauge. No TOPS classification.

85049   RH   393325 1956   CJ   c4/86   P   85049

used from new by British Railways at Chesterton Junction Permanent Way Materials Depot, Cambridge; to Northamptonshire Ironstone Railway Trust, Hunsbury Hill, 2nd August 1986; to Overland Railways, Chidham, near Chichester, for restoration, 1989; to Vobster Light Railway, Holwell Farm, Mells, Somerset, 13th January 1992; to Somerset & Avon Railway Company, Radstock, 25th June 1994; to Derbyshire Dales Narrow Gauge Railway, Rowsley, Derbyshire, February 1999.

85051   RH   404967 1957   CJ   c4/86   P   85051

used from new by British Railways at Chesterton Junction Permanent Way Materials Depot, Cambridge; to Cadeby Rectory, Market Bosworth, Leicestershire, 3rd July 1986; to Derbyshire Dales Narrow Gauge Railway, Rowsley, Derbyshire, 6th May 2006.

SECTION 26

Ruston & Hornsby Ltd built, 4-wheel diesel mechanical locomotive (Ruston's class 48DL). Fitted with a Ruston 4VRO engine developing 48bhp, and 1ft 6in diameter wheels. Ruston & Hornsby built 1,127 of this class, to forty different gauges, with works number 221615 (see Appendix C) being of 2ft 0in gauge and 224337 of 3ft 0in gauge. Neither locomotive was ever allocated a BR number. No TOPS classification.

-   RH   224337 1945   LSD   1964   P   06/22/6/2
used from new by British Railways at Lowestoft Sleeper Depot, Suffolk; to A. King & Sons Ltd, Norwich, September 1964; to Lynlite Concrete Co Ltd, Bury Road, Ramsey, date not known; disused by 1974; to J. & K. Harris, scrapyard, Norwood Road Industrial Estate, March, Cambridgeshire, December 1988; purchased by Andrew Wilson; to Green's Industrial Services, Sibthorpe, Nottinghamshire, for storage, 14th June 1995; to Andrew Wilson, Leeds, 27th March 1997.

SECTION 27

Ruston & Hornsby Ltd built, 0-6-0 diesel electric locomotives (Ruston's class 165DE) and numbered PWM650 to PWM654. Fitted with a Ruston 6VPH engine developing 155hp, and driving wheels of 3ft 2½in diameter. British Railways purchased five for Departmental work on the Western Region. Later re-numberered 97650 to 97654. No TOPS classification.

PWM650 RH   312990 1952   81D   4/87   P   PWM650/ 97650
97650   to Lincoln City Council, Holmes Yard, Lincoln, 27th February 1991; to Appleby Frodingham Railway Preservation Society, Scunthorpe, for storage, 4th February 1994; to Great Northern & East Lincolnshire Railway, Ludborough, by March 1995.

PWM651 RH   431758 1959   86A   9/98   P   97651
97651   to Northampton & Lamport Railway, Chapel Brampton, 10th November 1998; to Strathspey Railway, Aviemore, May 2008.

PWM653 RH   431760 1959   81D   ?   P   97653
97653   to Yorkshire Engine Company, Long Marston (despatched from BR Reading Depot), 6th November 1998; allocated YEC works number L163; acquired by John Payne from receivers of YEC, November 2001; locomotive remained at Long Marston; dismantled for spares; remains to Brian Hirst Recycling, Bullington Cross, Hampshire, 1st August 2011.

PWM654 RH   431761 1959   ?   ?   P   PWM654
97654   to Peak Rail, Rowsley (despatched from Slateford), 6th April 2005.

SECTION 28

Ruston & Hornsby Ltd built, 4-wheel diesel mechanical locomotive (Ruston's class 20DL). Fitted with a Ruston 2VSO engine developing 20hp, and wheels of 1ft 4½in diameter. Ruston & Hornsby built 1,198 of this class, to 37 different gauges, with works number 202005 being of 2ft 2in gauge. This locomotive was never allocated a British Railways number. No TOPS classification.

—-   **RH**   **202005 1940**   **HHC**   ?   **F**   —-
used by British Railways at Hall Hills Creosoting Depot, Boston; to John S. Allen & Sons Ltd, Upminster, by May 1967; locomotive's fate not known (it may have been exported, but confirmation is required).-

LOCOMOTIVE APPENDICES

APPENDIX A

A.R. Adams & Son, Robert Street, Newport, Gwent

This company hired locomotives to various concerns, mainly collieries and coal depots in South Wales. Between hirings the locomotives were stored and repaired at the nearby premises of United Wagon Company (Barker & Lovering), Newport, and at BR depots at Ebbw Junction (Newport) and Canton (Cardiff).

Details of known hirings are given below:-

D2139 to NCB Marketing Department, Gwent Coal Concentration Depot, Newport, (despatched from 85A Worcester Depot), 10th December 1968; to NCB Coal Products Division, Nantgarw, July 1969 to January 1970; to Coed Ely Coking Plant, March 1970; to Nantgarw Coking Plant, August 1970; sold to NCB Coed Ely Coking Plant, by 30th March 1971.

D2178 to NCB Aberaman Colliery, January 1970; to Wiggins Teape Ltd, Ely Paper Works, Cardiff, 24th February 1970; to Nantgarw Coking Plant, 21st August 1970; to Wiggins Teape (seen there 23rd August 1971); to Powell Duffryn Fuels Ltd, NCBOE Gwaun-cae-Gurwen Disposal Point, from about March 1972 to July 1972; sold to Coed Ely Coking Plant, by April 1974.

D2181 to NCB Marketing Department, Gwent Coal Concentration Depot, Newport (despatched from 85A Worcester Depot), 10th December 1968, and to whom the locomotive was sold by August 1971.

D2182 (moved to Adams on 29th November 1968); to NCB Coal Products Division, Caerphilly Tar Works, from about 29th November 1968 to February 1969; sold to Sir Lindsay Parkinson & Co Ltd, Glyn Neath, February 1969; re-purchased by Adams, about July 1972; to Lindley Plant, Gatewen, September 1973, to whom the locomotive was sold.

D2186 to NCB Aberaman Colliery, 8th February 1970 to about September 1970; to NCB Tower Colliery, from March 1971 to about September 1972; at Birds (Swansea) Ltd, Cardiff Docks, 23rd July 1972, to whom the locomotive was sold; scrapped, January 1981.

D2193 to Powell Duffryn Fuels Ltd, NCBOE Coed Bach Disposal Point, Kidwelly, from September 1969 to June 1970; to NCB Coal Products Division, Coed Ely, Tonyrefail, from July 1970 to about September 1970; to NCB Mountain Ash Colliery, from about February 1971 to about April 1971; to Monsanto Chemicals, September 1972; to NCB Coal Products Division, Nantgarw Coking Plant, by 31st May 1972 to early 1973; to NCB Taff Merthyr Colliery, from August 1973 to 7th December 1973; to NCB Garw Colliery, from October 1977 to 27th September 1978; returned to Adams; scrapped, January 1981.

D2244 to Monsanto Chemicals Ltd, Newport, Monmouthshire, from August 1970 to at least 30th March 1971; to NCB Coed Cae, Pencoed, by 28th March 1971 to about March 1972; to NCB Ogmore Central Washery, by 13th March 1972 to 19th May 1972; to NCB Gwent Coal Concentration Depot, Newport, by 18th July 1972 to December 1972; noted stored (with cab boarded up) at BR Pill, Newport, 21st December 1972; scrapped, January 1981.

12054 to NCB Mountain Ash Colliery (direct from 6C), from 15th September 1971 to May 1972; to NCB Tower Colliery, Hirwaun, from May 1972 to September 1973; to BR Canton Depot, Cardiff, for repairs, by 1st October 1973; to NCB Mardy Colliery, from 9th April 1974 to October 1975; at Ebbw Junction Depot, Newport, for repairs, on 11th October 1975; to NCB Mardy Colliery, by 16th March 1976 to September 1979; to BR Canton Depot, Cardiff, for repairs, September 1979; to NCB Mardy Colliery, from late November 1979 to January 1981; scrapped, April 1984.

APPENDIX B

T.J. Thomson & Son Ltd, Millfield Scrap Yard, Stockton-on-Tees

This company purchased three ex-BR Departmental locomotives, which were despatched to its yard at Stockton-on-Tees from BR Thornaby Depot in May 1970, having been stored at the latter location from about August 1969. They were never used at Thomson's works, but were stored for several years until scrapped in October 1981. The trio (which were all 4-wheel locomotives of the maker's class 88DS) were fitted with 88hp engines and diesel-mechanical transmission.

Departmental number	Builder	Works number	Year built	Last working location
56	RH	338424	1955	Etherley Tip, near Bishop Auckland
82	RH	425485	1959	Dinsdale Welded Rail Depot, near Darlington
87	RH	463152	1961	Geneva Yard, Darlington

APPENDIX C

Locomotives sold abroad

D2010 Swindon 1958 51L 11/74 P ?
03010 despatched from Tyne Yard; exported from Stranraer per Shipbreakers (Queenborough) Ltd, Kent; to Trieste, Italy, May 1976; disused by 1984.

D2019 Swindon 1958 32A 7/71 P 1
to Shipbreakers (Queenborough) Ltd, Kent, about June 1972; exported from Sheerness Docks to Stabilimento ISA, Ospitaletto, Brescia, Italy, September 1972; still in use, May 1997; disused by 30th August 1999.

D2032 Swindon 1958 32A 7/71 P 2
to Shipbreakers (Queenborough) Ltd, Kent, about June 1972; exported from Sheerness Docks to Stabilimento ISA, Ospitaletto, Brescia, Italy, August 1972; still in use, September 1991; extant but disused by 30th August 1999.

D2033 Swindon 1958 32A 12/71 P **PROFILATINAVE 2**
to Shipbreakers (Queenborough) Ltd, Kent, August 1972; exported from Sheerness Docks to Siderurgica, Montirone, Brescia Italy, August 1972; still in use, June 2003.

D2036 Swindon 1958 32A 12/71 P PROFILATINAVE 1
to Shipbreakers (Queenborough) Ltd, Kent, about June 1972; exported from Sheerness Docks to Stabilimento ISA, Ospitaletto, Brescia, Italy, August 1972; to Siderurgica, Montirone, Brescia, date not known; disused by 7th June 1998; dismantled and used for spares, by June 2003.

D2081 Doncaster 1960 31B 12/80 P 03081
03081 despatched by road from BR Swindon Works, 13th November 1981; exported to Sobemai, Maldegem, near Bruges, Belgium, November 1981; to Genappe sugar factory, Belgium, by 20th July 1991; extant but disused, 13th September 1999; returned to England; to Mangapps Farm Railway Museum, Burnham-on-Crouch, Essex, 8th March 2004.

D2098 Doncaster 1960 51A 11/75 P ?
03098 despatched from Tyne Yard; exported from Stranraer per Shipbreakers (Queenborough) Ltd, Kent; to Trieste, Italy, May 1976.

D2128 Swindon 1960 82A 7/76 P D2128
03128 to Bird's Commercial Motors Ltd, Long Marston, October 1976; resold for export and shipped from Harwich, December 1976; to Zeebouw-Zeezand (where re-engined and numbered 6G2), for Zeebrugge Port Authority harbour extension contract; at Gent depot (Belgium Railways) for wheel turning, March 1989; to Stoomcentrum, Maldegem, May 1989; returned to England (see main listing), 29th June 1993.

D2134 Swindon 1960 82A 7/76 P 03134
03134 to Bird's Commercial Motors Ltd, Long Marston, October 1976; resold for export, about January 1977; to Zeebouw-Zeezand (where re-engined and numbered 6G1), for Zeebrugge Port Authority harbour extension contract; at Gent depot (Belgium Railways) for wheel turning, March 1989; to Stoomcentrum, Maldegem, May 1989; returned to England (see main listing), 27th April 1995.

D2153 Swindon 1960 51L 11/75 P ?
03153 despatched from Tyne Yard; exported from Stranraer, per Shipbreakers (Queenborough) Ltd, Kent, to Trieste, Italy, May 1976; disused by 1984.

D2156 Swindon 1960 52A 11/75 P ?
03156 despatched from Tyne Yard; exported from Stranraer, per Shipbreakers (Queenborough) Ltd, Kent, to Trieste, Italy, May 1976; disused by 1984.

D2157 Swindon 1960 50C 12/75 F ?
03157 to Shipbreakers (Queenborough) Ltd, Kent, July 1976; exported from Sheerness Docks to Trieste, Italy, February 1977; rebuilt by IPE, Verona; to Chiari steelworks; scrapped on site, 23rd May 1997.

D2164 Swindon 1960 30A 1/76 F ?
03164 to Shipbreakers (Queenborough) Ltd, Kent, (despatched on 5th July 1976, arrived 16th July 1976); exported from Sheerness Docks to Trieste, Italy, February 1977; rebuilt at IPE, Verona; to Chiari steelworks; scrapped on site, 23rd May 1997.

D2216 DC 2539 1955 30A 5/71 P 3
 VF D265
to Shipbreakers (Queenborough) Ltd, Kent, 1972; exported from Sheerness Docks to

Stabilimento ISA, Ospitaletto, Brescia, Italy, September 1972; still in use April 1992; extant but disused (no engine), August 1999.

D2232   DC   2556   1956   52A   3/68   F   ?
         VF   D282
to Shipbreakers (Queenborough) Ltd, Kent; exported from Sheerness Docks to Attilio Rossi, Rome, Italy, August 1972; disused by 1984; scrapped, by May 1997.

D2289   DC   2669   1960   70D   9/71   P   NPT
         RSHD   8122
to Shipbreakers (Queenborough) Ltd, Kent, about April 1972; exported from Sheerness Docks to Acciaierie di Lonato, Brescia, Italy, April 1972; seen in use on 12th August 2005.

D2295   DC   2675   1960   70D   4/71   F   ?
         RSHD   8128
to Shipbreakers (Queenborough) Ltd, Kent, 18th March 1972; exported from Sheerness Docks to Acciaierie di Lonato, Brescia, Italy, September 1972; scrapped, 1982.

D2432   AB   459   1960   65A   12/68 F   ?
to Shipbreakers (Queenborough) Ltd, Kent; exported from Sheerness Docks to Trieste, Italy, March 1977; disused by April 1984; scrapped, 1986.

D2993   RH   480694 1962   70D   10/76 F   ?
07009   to Shipbreakers (Queenborough) Ltd, Kent, about March 1977; exported from Sheerness Docks to Trieste, Italy, March 1977; to Attilio Rossi, Rome, date not known; scrapped, by May 1997.

D3047   Derby   1954   70D   7/73   P   105
to BR Derby Works, January 1974; overhauled and modified; exported from Canada Dock, Liverpool, about 17th February 1975; to Lamco Mining Co, Tokadeh Mine, Nimba, Liberia, February 1975; last used 29th November 1981; extant, but derelict, May 2010.

D3092   Derby   1954   73C   10/72 P   101
to BR Derby Works, November 1972; overhauled and modified; exported (aboard the MV Avafors) from Middlesbrough Docks to Lamco Mining Co, Tokadeh Mine, Nimba, Liberia, May 1974; spare locomotive by 1986; last used 23rd February 1987; extant, but derelict, May 2010.

D3094   Derby   1954   73F   10/72 P   102
to BR Derby Works, March 1973; overhauled and modified; exported (aboard the MV Avafors) from Middlesbrough Docks to Lamco Mining Co, Tokadeh Mine, Nimba, Liberia, May 1974; last used 14th June 1980; used for spares, 1986; extant, but derelict, May 2010.

D3098   Derby   1955   73F   10/72 P   103
to BR Derby Works, March 1973; overhauled and modified; exported (aboard the MV Avafors) from Middlesbrough Docks to Lamco Mining Co, Tokadeh Mine, Nimba, Liberia, May 1974; last used pre-1980; used for spares, 1986; extant, but derelict, May 2010.

D3100   Derby   1955   75C   10/72 P   104
to BR Derby Works, March 1973; overhauled and modified; exported (aboard MV

Avafors) from Middlesbrough Docks to Lamco Mining Co, Tokadeh Mine, Nimba, Liberia, May 1974; last used 14th March 1982; used for spares, 1986; extant, but derelict, May 2010.

D3639   Darlington   1958   36A   7/69   F   ?
to C.F. Booth Ltd, Doncaster; to Conakry, Guinea, West Africa; exported from Surrey Dock, London, March 1970; used on the construction of the 135km Chemin de fer de Boke railway line, to take bauxite from mine to sea; dismantled by 1976; scrapped, early 1980s.

D3649   Darlington   1959   36A   7/69   F   ?
to C.F. Booth Ltd, Doncaster; to Conakry, Guinea, West Africa; exported from Surrey Dock, London, March 1970; used on the construction of the 135km Chemin de fer de Boke railway line, to take bauxite from mine to sea; dismantled by 1976; scrapped, early 1980s.

D9505   Swindon   1964   50B   4/68   F   98 88 202 2202-4
to Harwich Docks, 10th July 1975; to Sobemai, Maldegem, near Bruges, Belgium; exported from Harwich Docks; later sold to Suikergroep N.V. Opperstraat, Moerbeke-Waas sugar factory, near Gent; working on 22nd June 1997; scrapped on site, 1999.

D9515   Swindon   1964   50B   4/68   F   ?
to Spain; exported from Goole Docks, June 1982; stored at Madrid Charmartin since export, and still there 1988; scrapped, by October 2002.

D9534   Swindon   1965   50B   4/68   F   ?
to Harwich Docks, 10th July 1975; to Sobemai, Maldegem, near Bruges, Belgium; exported from Harwich Docks; sold to an industrial user near Milan, Italy, 1976; working at a steelworks near Brescia, by May 1997; scrapped, October 2005.

D9548   Swindon   1965   50B   4/68   F   93 71 131 0602-8
to Spain; exported from Goole Docks, June 1982; stored at Madrid Charmartin since export, and still there 1988; working at Sagrera yard, Barcelona, 19th June 1998, with RENFE track maintenance company NECSO EC name painted on its side; scrapped, by October 2002.

D9549   Swindon   1965   50B   4/68   F   P-601-03911-
                                                                            003-CMZ
to Spain; exported from Goole Docks, June 1982; stored at Madrid Charmartin since export, and still there 1988; at Zaragosa in October 2002; to Industrias Lopez Soriano, Calle de Miguel Servet, Zaragosa, by June 2003; scrapped March 2007.

LMS 7103   Derby   1941   ?   12/42   P   ?
withdrawn by LMS in December 1942 and sold to WD; used in Egypt; to Italy, April 1945; sold to Italian Railways (FS) in 1984; numbered FS 700.001; sold to a scrap merchant in Arquata Scrivia, near Genova; resold to Cariboni SPA, Colico, Italy; to Vercelli, Italy, by May 1995; to Piedmont Railway Museum, Turin, 1998.

LMS 7106   Derby   1941   ?   12/42   P   ?
withdrawn by LMS in December 1942 and sold to WD; used in North Africa in 1943, in Tunis by June 1943, in Algeria from November 1943 to March 1944; to Italy, March 1944; sold to Italian Railways (FS), 1946; numbered FS 700.003; withdrawn by FS in 1984; sold to a scrap merchant in Arquata Scrivia, near Genova; resold to La Ferroviaria Italiana Railway, Stia (who operate the branch line from Arezzo, west of Florence, Italy), 1991.

-   RH   221615 1943   MQ   ?   P   ?
2ft 0in gauge locomotive of the maker's class 48DL - for details see section 26; used by Southern Railway & British Railways at Meldon Quarry, Devon; exported to Egypt, date not known; derelict at Naga Hammadi Sugar Factory, Egypt, 1982.

1. D2037 working at NCBOE Oxcroft Disposal Point, Clowne, on 7th July 1994.
(Huw Williams)

2. D2078 preserved at the Stephenson Railway Museum, Chirton, on 12th April 2011.
(Adrian Booth)

3. D2112 working at Port of Boston, on 15th May 2001. (Huw Williams)

4. D2420 (06003) standing in the yard at UK Coal's Widdrington Disposal Point, on 23rd April 2006. (Huw Williams)

5. D2587 preserved at Peak Rail, Rowsley, on 10th May 2008. (Brian Cuttell)

6. D2858 shunting at Butterley Engineering Ltd, Ripley, on 26th March 1991.
(Huw Williams)

7. D2868 preserved at Peak Rail, Rowsley, on 24th May 2009. (Brian Cuttell)

8. D2953 preserved at Peak Rail, Rowsley, on 10th May 2008. (Brian Cuttell)

9. D2985 (HNRC, 07001) standing in the yard at Barrow Hill Engine Shed, Staveley, on 4th May 2011. (Adrian Booth)

10. D2987 (07003) standing in the yard at British Industrial Sand Ltd, Oakamoor, on 4th August 1982. (Huw Williams)

11. D2995 (07011) working at ICI Wilton Works, Middlesbrough, in July 1981.
(Terry Bye)

12. D2997 (07013) preserved at Peak Rail, Rowsley, on 24th September 2006.
(Brian Cuttell)

13. D3067 (08054) shunting at Tilcon Ltd of Grassington, on 14th May 2005.
(Huw Williams)

14. D3401 (08331) working at RFS (Engineering) Ltd of Doncaster, on 23rd March 1991.
(Andrew Smith)

15. D3689 (08527) standing in the yard at Flixborough Wharf, on 15th March 2011.
(Adrian Booth)

16. D3757 (08590) preserved at the Midland Railway, Butterley, on 30th April 2011.
(Adrian Booth)

17. D3836 (08669) working at Trafford Park Estates Ltd, Manchester, on 30th September 1993. (Huw Williams)

18. D3845 (08678) standing in the yard at Glaxochem Ltd of Ulverston, on 11th August 1989. (Andrew Smith)

19. D3898 (08730) awaiting repairs at LH Group, Barton under Needwood, on 21st August 2009. (Bob Darvill)

20. D3986 (HNRC, 08818) standing in the yard at Flixborough Wharf, on 15th March 2011. (Adrian Booth)

21. D4107 (09019) standing in the yard at Barrow Hill Engine Shed, Staveley, on 4th May 2011. (Adrian Booth)

22. D4141 (08911) standing in the South Yard at the National Railway Museum, York, on 1st March 2011. (Alex Betteney)

23. D4145 (08915) preserved at the Stephenson Railway Museum, Chirton, on 12th April 2011. (Adrian Booth)

24. D8568 (MR 9921 on left) 'on shed' at Hemel Hempstead Lightweight Concrete Co Ltd, on 17th March 1973. (Robert Pritchard)

25. D9517 standing in the yard at NCB Ashington Colliery, on 21st May 1977.
(Andrew Smith)

26. D9524 preserved at Peak Rail, Rowsley, on 10th July 2010. (Brian Cuttell)

27. D9537 working at Stewarts & Lloyds Minerals Ltd, Corby Quarries, on 8th November 1975. (Andrew Smith)

28. 12093 standing in the yard at NCBOE Widdrington Disposal Point, on 21st May 1977. (Andrew Smith)

29. ED10 (Ruston & Hornsby 411322) standing in the yard at Tracked Hovercraft Ltd, Earith, on 17th March 1973. (Robert Pritchard)

30. ZM32 (Ruston & Hornsby 416214) preserved at the Steeple Grange Light Railway, Wirksworth, on 29th August 2010. (Brian Cuttell)

31. 85051 (RH 404967, right) and 85049 (RH 393325, left) working at BR, Chesterton Junction PW Depot, on 17th March 1973. (Robert Pritchard)

32. Ruston & Hornsby 224337 standing in the yard at J. & K. Harris of March, on 21st April 1989. (Robert Pritchard)